SITE, SPACE, AND STRUCTURE

Kim W. Todd

 VAN NOSTRAND REINHOLD COMPANY
——————————————— **New York**

To Michael, in spite of his help;
to Loy, because of his;
and to Dane, who came after.

Copyright (©) 1985 by Van Nostrand Reinhold Company Inc.
Library of Congress Catalog Card Number
ISBN 0-442-28319-9

Printed in the United States of America
Designed by Loudan Enterprises

Published by Van Nostrand Reinhold Company Inc.
135 West 50th Street
New York, New York 10020

Van Nostrand Reinhold Company Limited
Molly Millars Lane
Wokingham, Berkshire RG11 2PY, England

Van Nostrand Reinhold
480 La Trobe Street
Melbourne, Victoria 3000, Australia

Macmillan of Canada
Division of Gage Publishing Limited
164 Commander Boulevard
Agincourt, Ontario M1S 3C7, Canada

16 15 14 13 12 11 10 9 8 7 6 5 4 3 2 1

Library of Congress Cataloging in Publication Data
Todd, Kim W., 1953-
    Site, space, and structure.

    Bibliography: p. 189
    Includes index.
    1. Building sites—Planning.   2. Architecture and
energy conservation.   3. Space (Architecture)   I. Title.
NA2540.5.T6   1985   720   84-15347
ISBN 0-442-28319-9

This book would have been impossible without the help of the
following people, whose assistance is gratefully appreciated:
Richard Austin, who kept me on track; Bud Dasenbrock, who
contributed facts and advice; Chris Beardslee, who drove and
drew; Jay Schluckebier, who listened; the secretarial staff of
Bailey/Polsky/Cada/Todd, who furnished copies, typewriter,
and ribbons; Maggie Breckel and Pete Jensen of the USDA Soil
Conservation Service, who ransacked the photo files; Lynn
Jones of Davis/Fenton/Stange/Darling, who let *me* ransack the
photo files; Ted Healey of Brown/Healey/Bock, who replied
quickly to my request for help; and of course Loy.

Contents

1 Process and Analysis

All site planning, whether as part of a large-scale community planning effort or for a single residential dwelling, involves developing the site, the structure, and the spaces that form the relationships between them in order to fulfill some human intention. The resulting design may be intensely active—with every square foot of space designed for constant activity—or it may be intended for visual use only, as a backdrop for activities on an adjacent property or as a sight corridor toward a landscape feature in the distance.

There is a difference between space that is planned and space that just exists. Planned space has a focus or orientation and is supportive of the activities that are to occur within it. Unplanned space appears forgotten or leftover. Consequently, because it does not utilize the site's potential as well as it could, unplanned space is a waste of the client's money: the land was purchased and should be used to the fullest extent possible.

The key to designing planned space rather than ending up with unplanned space lies in integrated design—consideration of all aspects of the project, including site, space, and structure, from the start. To attempt to design the structure independently of its site and of the spaces that define it is to invite unplanned space and an uncomfortable visual or physical relationship between built and unbuilt spaces. No matter what type of project is being attempted at a particular site, the land to be developed remains a constant. Even if the structure occupies the property from boundary to boundary, it will draw its character from the surrounding land and spaces, and neither site nor structure can exist without space. Thus, the site/space/structure relationship becomes the major determinant of the final design.

Land is continuous, whether it is visually fragmented by intrusive elements or site barriers or is proprietarily parceled by man-made "lines" such as property lines, easements, and rights-of-way. The challenge to the designer is to allow the land to continue, regardless of those barriers, and to be perceived and used in a smooth-flowing manner by means of subtle transitions between different areas of the site.[1]

The use of a logical process simplifies the design of site, space, and structure as an integrated whole, encouraging the designer to consider the three elements simultaneously and allowing the needed transitions to occur.

A process is an activity occurring over time that serves as a method for planning future activities. The use of a process is not exclusive to architecture and landscape architecture: it occurs in fiscal planning, in the development of marketing strategies for corporations, and (on a simpler scale) in the scheduling of a person's daily activities. The process in every case is only a symbol of the interaction between people and their environment; it is not the interaction itself. Processes allow people to catch the bus in time to get to work in the morning (the daily bus schedule serves as a reference for this process), to spend their time efficiently during meetings (an agenda serves the same purpose) or to play a football game (rules establish the procedural process).

Processes have been designated in many different ways and languages, depending on their intended use. Lawrence Halprin's "RSVP Cycles" designate an interaction among resources (R), score (S—how to work to get results), valuation (V—a decision-oriented analysis of the results of the score), and performance (P—what actually happens).[2] Several management models for

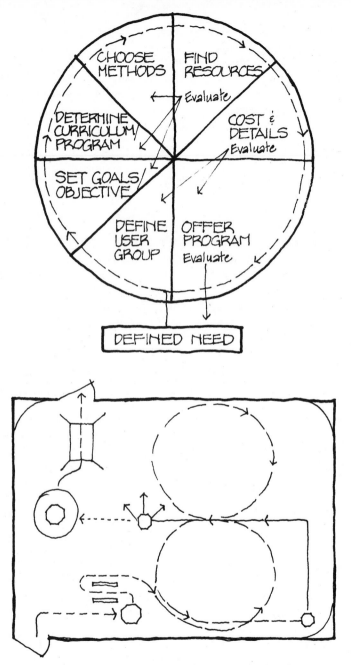

Inside diagram:

CHOOSE METHODS — FIND RESOURCES

Evaluate

DETERMINE CURRICULUM PROGRAM

COST & DETAILS

Evaluate

SET GOALS OBJECTIVE

DEFINE USER GROUP

OFFER PROGRAM

Evaluate

DEFINED NEED

1-1. Graphic representations of two different processes: a. an education program; b. a trial class program.

increased efficiency and productivity are based on a flow from input to output. A trail class in a horse show and the design of continuing education programs both use integrated processes (figure 1-1). Still, no matter how such processes are described, they have a number of elements in common.

First, any process (as understood here), no matter what its ultimate application, is cyclical in nature. This means that it may be entered or exited at any point, and reassessment of decisions and activities may occur at any time. Of course, in design as in most things, a point of diminishing returns can be reached, at which the cost of recycling may outweigh the potential benefits of doing so. The important point of this cyclical aspect of processes, though, is that feedback and revision can occur if new information comes to light or program requirements change.

Second, use of an integrated process forces the designer and client to consider the project in its entirety before focusing on the details. A design based on a clear understanding of the total scope of the project stands a better chance of evolving to meet the needs of the client than does a project developed on the basis of a single desired detail.

Third, the purpose of any process is to develop a plan, whether the plan is physical, as in most building designs, or procedural, as in the management of a corporation.

Most successful processes begin with a program that clearly defines the scope of the project. Programming is followed by analysis of existing conditions, development and choice of an acceptable alternative, implementation, and evaluation. The process may be very simple or highly complex, depending on the project. In some instances, the designer may merely be involved in preparation of construction documents, which is a part of implementation; in others, the designer's involvement may extend from detailed preparation of a program statement through postconstruction evaluation of the effectiveness of the design after several years. Most design work can be successfully accomplished if programming, analysis, alternatives, implementation, and evaluation are each addressed (figure 1-2).

Since the purpose of using a process is to guide interactions between people and the environment so that a desired relationship is established, the design process can be used at four different levels: land-use planning, comprehensive planning, master planning, and project planning.

LAND-USE PLANNING

Land-use planning determines the uses for which a piece of land is best suited, either independently of all factors other than the sense the land conveys of what it wants to be or based on situational need. Land-use plans are often regional in scope, developed for an entire county or along a significant length of a natural land feature such as a major river. A land-use plan is analagous to zoning in urban areas in that it establishes uses for certain site areas but does not address how those uses are to be implemented. For example, a county land-use plan may divide the site into agricultural areas, transportation corridors, existing and projected urban growth areas, recreation sites, and preserves. Once the land-use plan exists, a comprehensive plan may be developed for any one of the use areas (figure 1-3).

COMPREHENSIVE PLANNING

A comprehensive plan may encompass an area as large as a major city or as small as a privately owned hospital. This type of plan is a general overview of all site factors: traffic and circulation, space requirements, planted and open areas, buildings and other constructed features. A comprehensive plan establishes general locations, philosophies, and guidelines for major design features, but it does not go into specific detail. It may include, for example, a recommendation to eliminate vehicular traffic from certain areas (including general indications of which streets to close and what traffic patterns to accommodate on surrounding streets), designations of planting masses and of screening and street tree requirements, and even a general list of recommended plants. It will not, however, show the actual traffic corridor design or give the exact location and name of each plant on the site (figure 1-4).

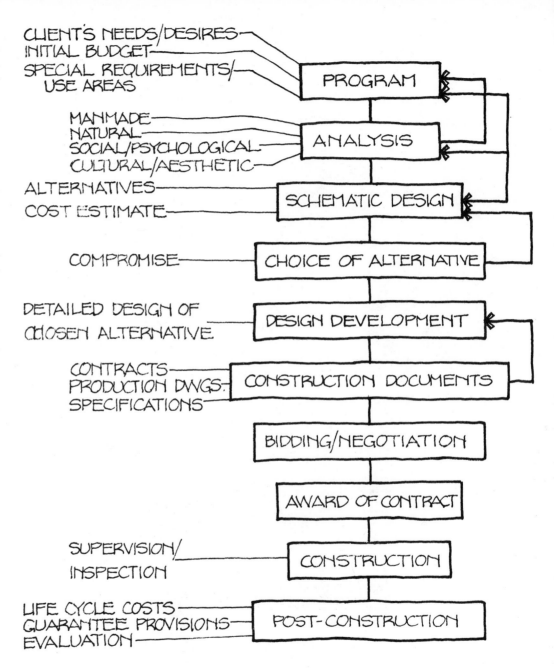

1-2. The design process.

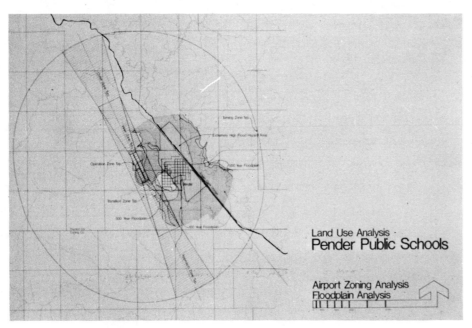

Land Use Analysis
Pender Public Schools

Airport Zoning Analysis
Floodplain Analysis

1-3. Land use analysis showing flight patterns and floodplains. Courtesy Davis/Fenton/Stange/Darling, Lincoln, Nebraska.

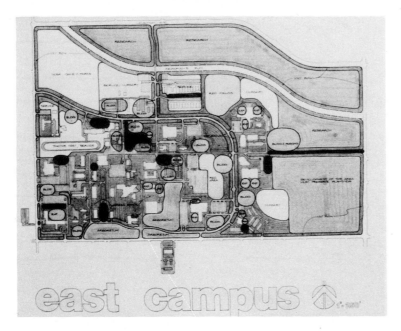

east campus

A comprehensive plan should be regarded as a framework for change. The development of such a plan should be based on a sensitive and thorough knowledge of the client, the project, and the goals of the institution or individuals who will control the project's future use. The designer should recognize, however, that a plan developed over a two- or three-year time span and intended to be a guide for the following five to twenty-five years must be adaptable to changing uses and needs. Successful comprehensive plans are designed to allow justifiable revision. The need for flexibility increases with the intended duration of the plan: a twenty-five-year comprehensive plan will eventually be administered by another generation of people, and different ideas and technical innovations will have been developed in the meantime that may suggest changes to the plan.

MASTER PLAN

A master plan is a more specific plan for developing one aspect of a comprehensive plan or for developing all aspects of a smaller site area. Master planting plans (which specify the composition of plant masses, either in terms of design characteristics or by genus name), master circulation plans (which set up a hierarchy of widths and locations, based on travel needs), and master development plans (which give the general location of such site features as ponds, trails, and seating areas) are all common applications of the master planning process.

A master plan is developed for a more immediate future than is a comprehensive plan. It is often presented to the client as a phased plan, the intention being that either certain features (such as all the paving) or certain areas (such as a playground) be developed in their entirety when the construction budget allows. In this manner, it also serves as a guideline for production of the construction documents according to which the project will be built (figure 1-5).

1-4. A comprehensive plan.

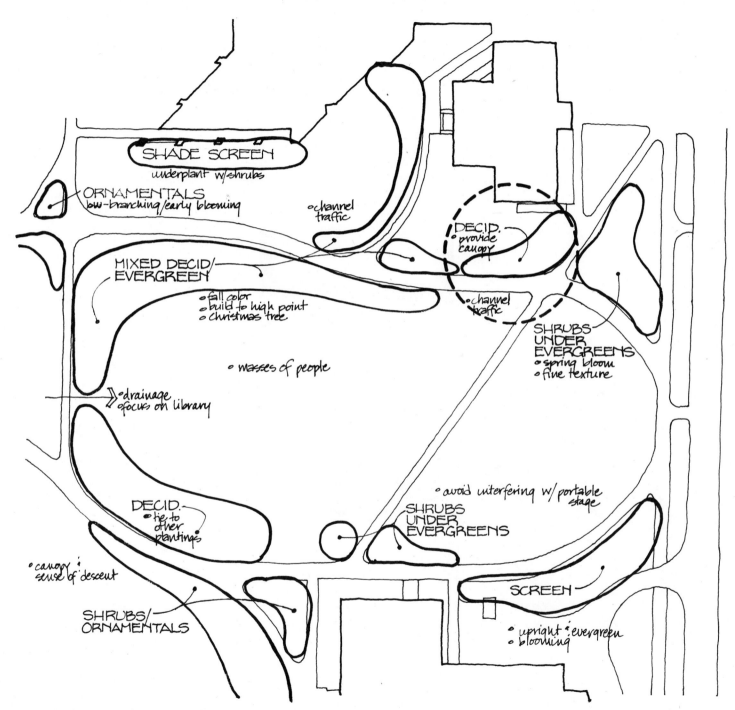

SHADE SCREEN
underplant w/shrubs

ORNAMENTALS
low-branching/early blooming

o channel
traffic

MIXED DECID/
EVERGREEN

o fall color
o build to high point
o christmas tree

DECID.
o provide
canopy

o channel
traffic

SHRUBS
UNDER
EVERGREENS
o spring bloom
o fine texture

o masses of people

o drainage
o focus on library

o avoid interfering w/ portable
stage

DECID.
o tie to
other
plantings

SHRUBS
UNDER
EVERGREENS

o canopy :
sense of descent

SHRUBS/
ORNAMENTALS

SCREEN

o upright : evergreen
o blooming

1-5. A master plan for an area of the comprehensive plan shown in figure 1-4.

PROJECT PLANNING

Project planning involves developing detailed schematics and design development drawings for a specific project. Project planning is short-term; often it is directly related to a time schedule based on project funding. All aspects of the project, from the precise location of the structure to the choosing and detailing of materials and specification of trees, are developed in enough detail to allow construction documents to be drawn (figure 1-6).

SITE SELECTION

The design process can also be used to help a client choose the most appropriate site for a project (if this is an option). The site selection process is used when a specific use is contemplated, but no predetermined site exists for that use. In this case, the process concentrates on analyzing the suitability of several possible sites for that particular project. All projects (except those in which a single structure occupies the entire property) involve site selection to some extent because several possible locations for activities may exist within the boundaries of a single site.

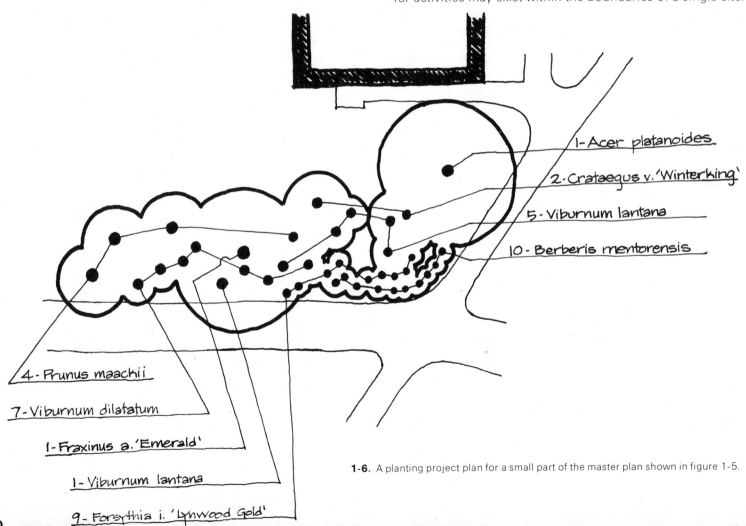

1 - Acer platanoides

2 - Crataegus v. 'Winterking'

5 - Viburnum lantana

10 - Berberis mentorensis

4 - Prunus maachii

7 - Viburnum dilatatum

1 - Fraxinus a. 'Emerald'

1 - Viburnum lantana

9 - Forsythia i. 'Lynwood Gold'

1-6. A planting project plan for a small part of the master plan shown in figure 1-5.

The design process consists of a series of steps or alternatives through which the designer proceeds in an orderly and logical fashion in order to produce the most suitable design for a given situation. The primary steps (which are themselves divisible into parts) are development of a program, analysis of site data, schematic design, presentation to and choice by the client, design development, drafting of construction documents, bidding or negotiation of contracts for site work, and construction.

PROGRAM

The program is the basis for all subsequent design decisions. Programming defines the needs and desires of the client, and uses these as guides in developing satisfactory use areas. The program statement should identify any special requirements of the project associated with the land or easements on the land, provide a preliminary budget statement, and project the time schedules for design and construction.

Needs and Desires

The successful completion of a project depends upon continual reference by the designer to the program to make sure that the client's actual needs are being met by the design and that the client's needs and desires are understood in the same terms by both client and designer. It is only through the input of the user and/or client that these needs and desires can be accurately identified.

If the client remains in the background, the designer will be forced to try to interpret what is wanted in the project from a few original, ambiguous signs. This can also happen if the nominal client and the actual users are different people. The staff, maintenance and service people, delivery people, and visitors fall into this second group. The designer is likely to create a far more sensitive and functional design if the input of those people is solicited during development of the program statement. The more conversations the designer holds with different users, in different settings, under different circumstances of formality, and at different times, the more information will come to light. People talk about a project differently when they are on their home turf than when they are in an office; much can be revealed in casual conversation or by watching the ways people live and work.

Clients may come to the designer with a vague notion of needing something that they are not able to define exactly. The designer should then help them define their needs in a clear, concise written program. Difficulties can also arise when what the client *says* is needed and what actually *is* needed are two different things. For example, the client may claim to need parking spaces for one hundred cars, while zoning requirements and projections of staff and visitor use may indicate that only eighty spaces are needed. The rest of the spaces, therefore, are only *desired*.

In situations such as this, all of the designer's tact and professionalism come into play as he or she attempts to make the client understand the cost of those desired spaces. Perhaps the site and budget are both large enough to accommodate the additional twenty spaces; if so, the conflict between real and perceived need is not critical. If on the other hand the site cannot handle the larger number, the designer must rely on what had better be a good working relationship with the client to work out the problem.

ANALYSIS

To analyze is to separate a whole into simpler components and to understand each of those components on an individual basis before attempting to understand them in relationship to one another and to the whole. Analysis is as important as programming in guiding the development of the site plan. Analysis is never complete; prior to each design decision, the designer should consider the overall effect on existing conditions that will be produced by that decision (figure 1-7).

Analysis can easily become a self-perpetuating monster, with data begetting data until the designer has lost any sense of the reason for starting the exercise in the first place. Conversely, analysis can be oversimplified to such a degree that it is virtually worthless as a design tool. One of the most difficult aspects of site analysis to learn is to distinguish between irrelevant factors and those that have a potentially significant impact on the project.

An example of an often-ignored factor that may have a great impact on the final design is the colors of the surrounding structures—the feeling of the designer often being that, since they cannot be changed, they need not be considered. However,

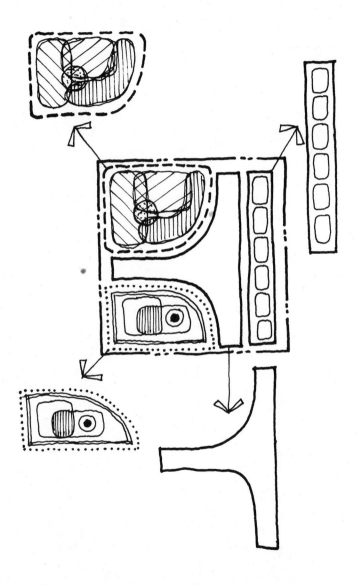

the presence of those colors outside the designer's control makes them a determinant in choosing nonclashing colors for the structure itself.

On the other hand, once the designer has reached the level of site-specific analysis, it is frequently unwise to base the analysis on lists of region-wide average temperatures for each month, average degree-days, and average hours of sunshine. Instead, the particular shadow patterns and microclimate variations that occur on that site as a result of those climatic conditions should be assessed.

Analysis consists of two basic components. The first is site observation—gathering all pertinent facts from whatever sources of information are available. The second is the interpretation or assessment of each piece of data, either to establish its relevance to the project or to discard it.

Sources of Information

There are many sources of information, but the best are the designer's own senses. Personal reconnaissance of the site and environs allows collection of much specific information about existing features that is unavailable from any other source. For example, the site survey may show the locations and even the diameters and common names of all trees, but only a personal visit will reveal whether a particular tree is tall and sparse or has a symmetrical form or is weak-wooded and declining. The same can be said for off-site features. Rarely will a survey extend beyond the boundaries of the property, and even if it does, it will tell the designer nothing about the color, height, and degree of maintenance of the surrounding structures.

It is also only through the on-site visit that the quality of the site can be determined. Each site has its own unique character, arising from such things as the quality and quantity of light reaching the site, the level and variety of sounds, the direction and speed at which one approaches and enters the site, and the influence of surrounding features (including buildings and trees). None of these can be adequately analyzed from a plan or even from photographs. The designer must personally experience the site to assess the full impact of each qualitative factor. Designs that consider and take advantage of those unique features of a site are memorable and succeed in setting the site and structure apart from other, similar developments (figure 1-8).

1-7. Analysis—separation of a complex element into simpler components.

Past property owners and users can also provide worthwhile information for site analysis. They may know of buried foundations, unusual soil problems, or infrequently occurring but annoying off-site influences such as a twice-yearly emptying of adjacent sludge ponds. It is important to contact the users as well as the owners because those who maintain a site and are present on it on a day-to-day basis usually have a good feel for its character and for the small but potentially serious problems that need to be defused during the initial stages of design.

A wealth of information commonly is available in the form of plans, surveys, and maps. To augment this written information, a current survey of the property is invaluable in accurately establishing the locations of existing features.

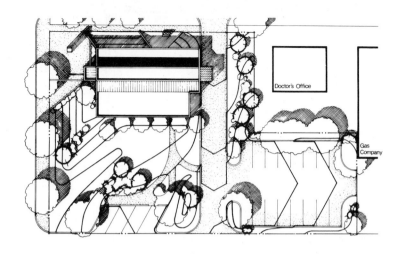

Site Plan **Falls City Federal Savings & Loan**
Auburn, Nebraska

a.

1-8. Plan information does not always show the quality and character of the site: a. site plan; b. actual developed site. Courtesy Davis/Fenton/Stange/Darling, Lincoln, Nebraska.

b.

1-9. A combination property-topographic survey.

Types of Surveys

Two types of surveys are commonly used in analysis: the property survey and the topographic survey. The property survey is a legal document that identifies the exact location, length, and angle of each of the property lines, the easements and other restrictions on the property, the rights-of-way and setback lines, applicable zoning restrictions, adjacent property owners, and the location by dimension of all structures and other constructed features, all trees, and all natural elements, as requested. The property survey also provides a written legal description of the property and (often) a regional location map.

The topographic survey identifies the existing elevations of the land and structures on the site. The degree of accuracy involved depends on the scale of the site and the scope of the project. On small sites, elevations may be given on a grid laid out in increments of ten feet or less, whereas a one-hundred-foot interval is common on large sites. In addition to the grid points (the points of intersection of the incremental lines of the grid), elevations may be given for existing trees whose diameters exceed a certain minimum size, for high and low points off the grid, at the tops and bottoms of curbs, walls, and stairs, and for the grate or invert elevations and flow lines of storm sewer systems. In addition, topographic surveys identify the depth of the cover over other underground utilities and the names, diameters, and spreads of all trees and plants.

Both types of surveys include the stamp of the surveyor, the date, a statement attesting to the accuracy of the survey, the benchmark(s) used in establishing it, a north arrow, the scale used, and the title of the project.

Both property and topographic surveys are based on a series of benchmarks. A benchmark, or datum, is a permanent, accurately described, and identifiable reference point that other surveyors and contractors can use to establish the same dimensions and elevations at a later time. Benchmarks are commonly expressed in feet above sea level, the form used by the United States Geological Survey (USGS). A list of the benchmarks established and recorded in an area is available through the USGS.

Sometimes the listed benchmarks may not be convenient to use on a given site because of problems of distance or visual interference with a clear sight line. In such cases, a set of benchmarks may be arbitrarily established. However, the same rules apply to them as to recorded benchmarks: they must be permanent (at least for the duration of the project), and they must be accurately described and identifiable. All benchmarks, whether permanent or temporary, are noted in their exact location on the plan, and listed in the legend.

The owner, rather than the designer, contracts for the survey; this enables the designer to avoid possible liability problems involving the surveyor's competence. However, the request for type, amount, and detail of information is made to the owner by the designer. Usually, a combination property/topographic survey is requested to obtain all information essential for design (figure 1-9).

Other Resources

Aside from the surveys, other types of written resources may be available. In many cases a project is required to comply with conditions established in a comprehensive plan. Referring to the comprehensive plan during analysis informs the designer of proposed changes that might affect the site in the future. There may also be natural restraints on site use, such as floodplains or ecologically unstable areas. Present and projected traffic counts for the area surrounding the site can suggest approaches to future traffic corridor development. Zoning changes may have been proposed for adjacent sites.

Computer-assisted analysis, used with increasing frequency in assessing developments of all scales, can provide satellite imagery of vegetative cover, soil types, and water- and drainage ways. Aerial photographs of the site taken in different seasons are useful in determining changes in vegetative cover, precipitation patterns, and drainage problems; these are particularly important points, since the analysis phase of most projects does not last through a complete change of seasons.

Many public agencies are excellent sources of information. Utility companies can provide accurate locations of under-ground and overhead lines, both existing and proposed; the Soil Conservation Service can supply information about soils and geologic conditions in the area, about the heights of water tables, and about the types of plants commonly found on certain soils in certain areas (especially useful on low-maintenance projects where the use of native plant materials is critical to the success of the project); and planning and zoning commissions can offer their knowledge of setbacks, building restrictions, zoning changes, and planned development.

The importance of gathering information about proposed future conditions either on or affecting the site cannot be overstated. Prior knowledge of a potential disruption allows it to be dealt with before rather than after the fact; thus, if the designer knows that an adjacent property is to be developed as a major industrial park within a few years, steps can be taken immediately to screen the site from offending noises, views, and odors and to maintain the site as its own design area.

Site analysis is all too often confined to the limits of the property lines, as though off-site factors were utterly uncontrollable. Designers must realize that, although they can rarely change anything outside the boundaries of the property, they can design the site to respond to conditions around it. A small site can be manipulated to take advantage of a view of a park or a landscape feature several blocks away; an unsightly view of the property can be avoided by plantings designed to direct the eye away from the offending element.

Interpretation of Information

Once the initial site observation is completed, interpretation or assessment of the information begins. Since the purpose of analysis is to guide later design efforts, each bit of information should have some relevance to the project and should in some way affect the design. Thus, every statement made in the analysis should be reflected in later design decisions. The interpretative phase of analysis can make the justification of future design decisions easy, by explaining *why* each analytical observation was made.

For example, the notation "Cold north wind" gives the client no sense of how the wind should affect formulation of the final design; nor does a label such as "Pedestrian path" accompanying a series of footprints on the base map convey the importance of this fact. However, by adding to the statement "Cold north wind" a key phrase such as "causes harsh walking environment and increases heating load on building," the designer will remember when referring to the analysis that a solution for this problem needs to be devised; at the same time, the client can be more easily convinced of the importance of any design decisions justified by such an interpretation.

The analytical observation should be as objective as possible to avoid causing the design to take form too early in the process. Analysis forms a part of problem identification leading to the solution, but it is not in itself the solution. An example of a design solution for the above statement might be: "Cold north wind should be blocked by mixing deciduous and evergreen plants in a windbreak to reduce heating load on building." At this early stage, however, the designer does not know that a windbreak composed of plants is the right decision; later information may reveal that the site for the major outdoor activities should be moved or that there is only room for a vertical screen. In other words, each statement important enough to include in the analysis should describe completely but nonprescriptively the condition or object, and should state its significance to the project in the same terms.

Another caution regarding the interpretation of analytical information involves the use of subjective words such as "good" and "ugly": while such words may accurately express the designer's sentiments, they should be avoided at least until a clear sense of the client's feelings can be obtained; this will reduce the possibility of unintentional misunderstandings. For all the designer knows, the client may have made a fortune on that "ugly" railroad spur adjacent to the building site.

Organization of Analytical Information

The voluminous information that is often a part of site analysis can make organizing the analysis difficult. One method of organization divides the information into four categories: *man-made or manufactured* elements, *natural* elements, *social-psychological-cultural* elements, and *qualitative or aesthetic* factors. This system can be used for both on-site and off-site features; some overlap among categories is unavoidable. Using a checklist within each of the categories can help the designer avoid overlooking something critical. Following is a list of many of the items that should be assessed during analysis.

Man-made or Manufactured Factors
  Circulation
    Existing systems within the site
    Existing systems available to and from the site
    Service access
    Orientation of the systems: was the site considered when they were installed?
    Relationships between systems
    Variable traffic numbers using the systems: can this variability be documented, or do the systems function randomly?
    Constraints against changing the existing systems to accommodate your needs
    Volume of noise generated by the systems
    Controls on the systems (traffic signals)
    Compatibility of the systems with the existing grade
    Difficulty/ease with which each system can be accessed from the others
    Level of maintenance currently practiced and needed
    Relationship of parking areas to systems
    Walking distances
    Provisions along the way to provide interest, focus attention, and distract viewer from undesirable features
    Scale/relative dominance of the systems on site and on the approach to the site
    Visual relationship of approaches to site
  Utilities
    On- and off-site locations of critical utilities
    Distance to hookups
    Visual intrusion by utilities
    Specific site problems caused by utilities, such as soil burnout from steam tunnels
    Accessibility of utilities for repairs
  Zoning Requirements
    Zone of site and surrounding areas: will the two, if different, be compatible?
    Easements that may affect building location/dimensions
    Setbacks; height and materials requirements or limitations
  Structures and Paving
    Present structures on and off site and their probable effect on the site's functional and visual quality

Blockage of views or solar access by adjacent structures

Climatic effects of surrounding structures and manufactured surfacing

Visual intrusion on site by surrounding structures, either because of their appearance or because of a view they have into the site

Degree of maintenance provided to and required by existing materials

Pollutants

Visual and olfactory pollutants present

Sources of pollutants and possibility of doing something about them

Daily or seasonal variability

Natural Factors

Sun and Shade

Quality and quantity reaching site

Direction

Effects of surrounding structures on sun and shade

Density of shadows cast

Changes in appearance, feeling, and function of site depending on time of day or night

Glare problems

How sun and shade will affect the functioning of the circulation systems

Wind

Microclimatic effects of wind: tunnels, dead spots, eddies

Seasonal changes of wind

Odors or trash and debris carried by wind

Blockage and direction changes caused by adjacent site conditions

Wind erosion signs

Possible structural or functional problems caused by wind, such as difficult-to-handle entry doors

Temperature

Hot or cold spots created by the interactions of other climatic factors

Variations in temperature from area to area

Water and Precipitation

Effects of surrounding conditions on precipitation: blocking it, changing its direction or intensity, creating dry areas

Snow and ice buildup areas

Fog patches

Ground water visible on the site, in the form of low spots or actual bodies of water

Subsurface water conditions, including water table, underground streams, aquifers

Surface drainage conditions, including surfacing that water flows over and eroded or poorly drained areas

Existence of man-made features to handle runoff

Problems from adjacent sites adding to water conditions on site

Quality of water

Maintenance problems associated with water conditions on site

Vegetation

Type, amount, and quality of existing vegetation

Vegetation on surrounding sites and its effect on the proposed project

State of maturity of plantings

Variety of plantings

Design characteristics and their suitability to the proposed project

Degree of maintenance needed and given for plants

Effects of vegetation on climate

Effects on visual appearance and feeling of the site

Seasonal change

Sensual appeal—smell, touch, sight, sound

Possibility of relocating vegetation to other parts of the site

Scale and aesthetic appearance

Wildlife potential or problems associated with plantings

Potential for plants to damage structures or paved areas—root intrusion, dropping branches

Wildlife

Evidence of beneficial or harmful wildlife

Possible site conditions that would attract wildlife

Soils

Suitability for structural support, supporting plant materials

Type and condition: clay, sand, loam; heavy or light, compacted or porous

Changes in soil type across the site
Acidity or alkalinity
Topsoil present on site
Topography
  Steepness or flatness
  Uniformity
  Relationship to surrounding grades
  Existing elements whose grade cannot be changed—
    places where new grading must meet existing features
  Erosion
  Orientation of slopes
Social/Psychological/Cultural Factors
  Attitude toward Environment (evident from the degree of de-
    velopment and maintenance of site and area)
  Social Influences
    Accepted way of designing and caring for a site in the
      area
    Traditions such as a showpiece front yard
  Sociability of Site and Surroundings
    Openness or private feeling
  Sensory Perception
    The feeling of the site: inviting, forbidding, open, intimate
    How the feeling changes with time of day and seasons
    Effects of the surrounding elements on perceptions
    Feelings of anticipation or foreboding generated on the
      visual and physical approaches
    Effects of color on perceptions
  Scale
    Scale of the site in relation to its surroundings
    Scale of site features in relation to the site itself
    Places where the site could "borrow" from other areas to
      increase or decrease scale
    Dominance of spaces or human beings
  Balance
    Dynamic or static balance within the site
    Balance of elements on site
    Balance of the site with surroundings
Quality and Aesthetic Factors
  Views and Vistas
    To the site and from the site, from all different angles of
      approach

Changes caused by sun and shade changes or by climatic
  conditions
Predominant site features toward which a good view or
  vista could be oriented
Need for buffering or screening views
Degree to which views are sequential
Changes in view and vista that take place with mode of
  transportation used
Level at which best views are attained: eye level, ground
  level, or higher?
Backdrop and foreground for the site
Form and Shape
  Form taken by the open and enclosed spaces on the site
    and around it
  Individual forms of interest that could be used as design
    features
  Relationships between forms
  Repetition or variation in forms

A thorough analysis of all of these factors may not be required for a given project, and the emphasis of the analysis will vary depending on the intended uses of the site. For example, if a site containing a number of older homes is to be cleared to make way for an apartment complex, the houses will have importance only in their various aspects as elements to be removed: their number, their sites, the amounts of salvage they contain, and their locations so that basements and foundations can be removed and utility lines cut. If, however, the area is to be restored as a historic district, then the condition of the houses—their size, shape, color, materials, maintenance, and orientation, the plant materials around them, and their proximity to one another—becomes important, since each of these factors will influence the outcome of the overall site plan. One of the first lessons a designer must learn, therefore, is how to assess the importance of the analytical factors for a given site (as well as how to weed out irrelevant information before a great deal of time is spent wondering how to make it useful).

After the various factors involved in a particular project have been identified and categorized, the visual organization of the factors by category remains to be done. Three different ways of handling this step are to categorize all the factors on one sheet, to key the factors directly to the plan, and to use overlays for each category (figure 1-10).

Site analysis is generally tied directly to the project program. Information about preexisting conditions is examined differently for a specific site with a specific use than it would be for a site lacking a preconceived use. If the program statement changes during the course of project development, interpretation of the analytical information may also change to reflect different uses or the greater or lesser importance of certain information. This project-specific analysis is subjective in nature: the designer knows that a specific site and program must be developed and that all work must be planned to proceed within the potential of the site. Nonspecific analysis can be much more objective, with a designer looking at a site simply for what is there and not for what is there with an eye to what must eventually be there. This relates directly to the previous discussion of site analysis, site selection, and land use analysis.

Once analysis and interpretation of the information gathered during analysis are (at least for the time being) complete, the actual design work begins.

SCHEMATIC DESIGN

The main goal of schematic design is to develop as many viable yet distinctly different alternatives as possible. During schematic design, creativity comes to the fore without being overly expensive to either the client or the designer (figure 1-11).

There are three key points to keep in mind during schematic design:

1. The alternatives should be creative and as distinctly different as possible, yet realistic. Designers sometimes find that an alternative presented almost as a joke to the client because of some outlandish technique or impossible-to-construct feature is just the one the client wants, and the designer is then forced to tell the client that it will not work.
2. The alternatives must respond to the conditions of the site as determined in the collection and interpretation of data during analysis.
3. The alternatives must meet the program requirements and thus the needs of the client.

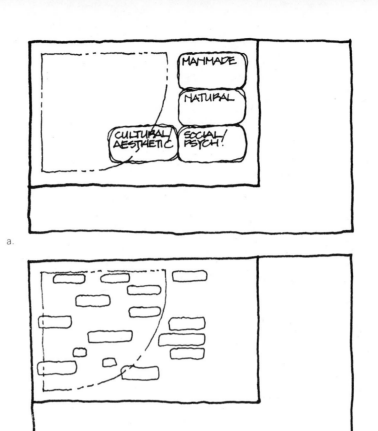

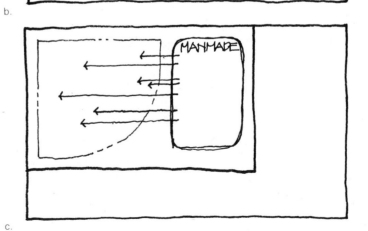

1-10. Three methods of organizing an analysis: a. categorized on one sheet; b. keyed directly to the plan; c. using overlays.

Schematic design can be approached in two different ways. The first is whole-site design—designing for all parts of the site with the intention of meeting all program requirements by means of a single solution, before going on to alternative solutions that again address all issues of the site at once. The whole-site ap-

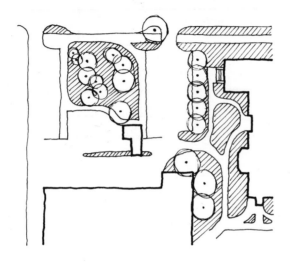

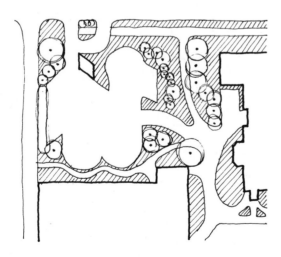

1-11. Two schematics for the same site and program.

proach is regularly used in design work, in the form of bubble diagrams that show relationships between site areas.

The second is fragmented design—developing as many alternatives as possible for only one aspect of the site plan or one requirement of the program (such as the layout of the buildings, or the play area, or the grading scheme), before going on to do the same thing individually for the remaining aspects of the site. An alternative for each fragment of the design is then paired in as many workable combinations as possible with alternatives for every other fragment.

There are advantages and disadvantages to both approaches. A significant advantage of whole-site design is that the designer is not tempted to go beyond schematic design into design development, but rather keeps the entire design loose and conceptual. This encourages the designer to think in terms of the whole site and the relationships between various parts of the design. A disadvantage of this type of design is that the mere idea of creating an alternative for the entire site all at once can be overwhelming, particularly since changing one site feature often changes all the relationships in one or more other site areas.

The advantage of the fragmented approach is that the designer can concentrate on solutions for a single aspect of the design, without continually trying to anticipate the "what if" situations associated with changing one portion. A disadvantage of this approach is that the designer can easily go beyond schematics and into specific detail, which then makes putting the pieces of the whole site back together again considerably more difficult. This can lead to a design solution that blends together poorly.

Another important part of schematics is the cost estimate. At this early stage, it will still be very rough, but it should be accurate enough to compare and contrast the one or two alternatives that best meet the requirements of the program and yet remain within the budget.

Developing a variety of alternatives is essential in schematic design. The greater the variation between alternatives, the easier becomes the selection of one at the next stage in the design process. During schematic design, a few solutions—either for the entire site or for certain elements of the design (for example, the physical relationship of the entrance roads to the service docks and parking areas)—are likely to recur repeatedly. Such recurrent solutions can be interpreted by the designer in one of two ways: as having significant appeal from a design standpoint, or as representing the most functional relationship between site areas or activities. In either case (or because of both), the recurrent solution provides a starting point for eliminating some of the schematics from consideration.

The designer may eliminate all but one of the solutions that show the recurring feature and may offer one or two significantly different features for presentation in alternative solutions: Since design is subjective and personal (the client hires a designer on the basis of personal reputation or because of some well-liked design work), the designer is likely to find one or two of the alternatives more appealing than the rest. The preferred alternative, obviously, will be easier to sell to the client than some of the others.

PRESENTATION AND CHOICE

Generally, the schematic design stage is followed by presentation to the client of two or more alternatives. At this point, the variations between alternatives become important. Presentation of a few alternatives with rather marked differences gives a client a clear choice, while presentation of alternatives involving subtle changes of form or placement, when presented to a client, is more likely to confuse than spark enthusiasm. The designer should also avoid presenting every single alternative; too much choice can be worse than too little.

The designer helps the client choose a solution to develop further, justifying the design on the basis of the program, the analysis and interpretation of the analysis, and the cost estimate figures. The final choice may be one of the alternatives exactly as developed, but it is more likely to be a combination of the best features from several of the alternatives (figure 1-12). The designer then takes that alternative into design development, the next stage of the design process.

1-12. Choice of a solution involving compromise.

DESIGN DEVELOPMENT

Once design development begins, it becomes more and more costly to do an about-face in the design and start over on a new concept. Instead, the changes at the design development stage should be minimal, consisting of minor adjustments in alignment or grade and in choices of materials or details.

The design development drawings are carried to a point at which a realistic and fairly precise assessment of the workability and cost of the design can be determined. All site features are located exactly as they will be situated when constructed; they are drawn true to scale. The bubbles indicating plant material masses, open spaces, or even building clusters assume their final form, with plants located as desired, outlines of buildings and other structures made realistic, and roads drawn to their correct widths.

The cost estimates associated with design development are based on accurate materials specifications. For example, a specification may be made for herringbone-pattern brick laid on a two-inch sand bed, whereas in the schematic estimate for the same thing the specified item may simply have been brick pavement. Accurate quantity takeoffs are also done as a part of design development, to improve the accuracy of the cost estimate further.

The presentation of design development drawings to the client is more critical and time-consuming than is the presentation of schematics. Any substantial changes made after this point will be costly to both the client and the designer and are likely to set back the time schedule of the project, so both the designer and the client should go over the design development drawings with extreme care.

CONSTRUCTION DOCUMENTS

Approval of the design development drawings signals the go-ahead for construction documents. The construction documents consist of the production drawings or blueprints, the written specifications of materials and work standards, and the contracts.

Production Drawings

The production drawings are drafted as accurately as possible, with all site features dimensioned in such a way that the contractors following them will be able to proceed with their work in logical fashion (figure 1-13). Depending on the scope of the project, the drawings for a site plan may consist of anything from a single sheet with an attached paragraph or two of specifications to ten or more full-size prints and a bound volume of specifications.

Production drawings form the basis for bids from contractors; inaccuracies in the drawings on the part of the designer can result in faulty bids. Both existing and proposed conditions should be shown on production drawings in order for the bidding contractors to determine the amount of work involved. Showing only the proposed conditions, especially of such site features as grading and drainage, forces contractors to guess (from the plans and/or a site visit) at the change between existing and proposed conditions; an estimate based on such a guess almost invariably goes in the contractor's favor.

A set of production drawings commonly consists of eight types of drawings: a location map, a demolition plan, a site layout plan, a site grading plan, a site drainage plan, a site utilities plan, a site planting plan, and detail sheets.

Location Map

The location map shows the position of the site in its surroundings. It is particularly useful as a means of identifying the major travel routes to the site.

Demolition Plan

The demolition plan shows all existing site features that are to be removed, identifies who (contractor or owner) will perform the removal, and states what (proper disposal or salvage) will be the ultimate disposition of the removed material. Demolition plans also include notes about protecting the features that are to remain (particularly if they are in close proximity to something being removed) and about shifting the position of any site features that are to be relocated on the site.

Site Layout Plan

The site layout plan shows the exact location of all proposed construction, including structures, pavement, retaining walls, benches, and other site amenities. A combination of angles and dimension lines is used to enable the contractor to locate each feature accurately. A common, permanent reference line—such as a property line or the edge of a preexisting sidewalk that is to remain—is used as the basis for the dimensioning system. Each feature is located as simply as possible, preferably by a single angle and a single dimension. The dimensioning sequence is based on a logical sequence of work the contractor is likely to follow.

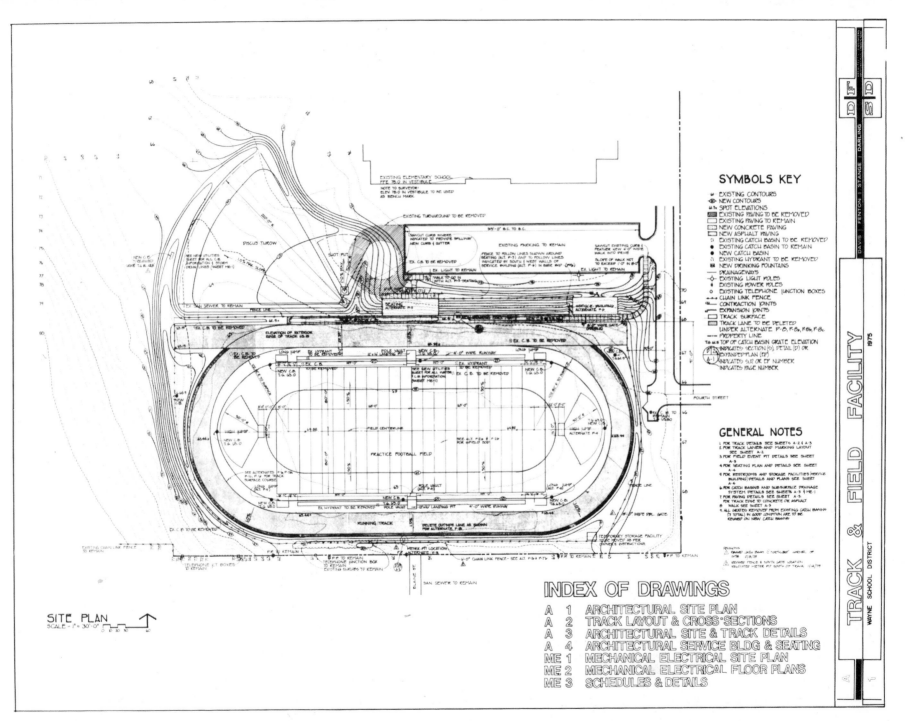

1-13. Site construction drawing. Courtesy Davis/Fenton/Stange/Darling, Lincoln, Nebraska.

Site Grading Plan

The site grading plan shows both existing and proposed contours of the site (to allow the contractor to make accurate earthwork takeoffs), spot elevations (at all corners, curbs, steps, walls, and other vertical variations), high points and low points, the location and size of all underground storm sewer pipe (and its type of construction), the rim and invert elevations at all manholes and points of connection between pipe, and the location of all culverts and headwalls.

Site Utilities Plan

Depending on the complexity of the grading plans and the utilities plans, these may or may not be combined. The site utilities plan shows the location and depth of all electrical, phone, steam, water, and gas lines, as well as their connections to existing lines. Locations of manholes and light poles are also provided; the underground irrigation system may be shown on the utilities plan, or it may require a separate sheet.

Site Planting Plan

If the site grading plan is too complex to allow inclusion of planting information, a separate planting plan must be drawn. The site planting plan provides accurate locations of all trees, shrubs, groundcover, flowers, and bulbs. It also specifies turf areas, mowing strips or special treatments for the edges of planting beds, and special care instructions for existing materials that are to remain. The plant schedule may be included on the sheet or with the specifications. It lists each plant by common and scientific name, states the size and conditions of purchase (bare-root, balled and burlapped, or in containers), identifies the planting season, and spells out any special requirements.

Detail Sheets

Details for all different types of construction and all construction materials are drawn either on separate sheets or in blank spaces on the plans. Details may include such disparate items as sections through various types of pavement (showing how one pavement joins another), wall and stair details, and planting details. Details are particularly important when the designer considers it critical to use certain materials or to construct site features in a certain way that may deviate from standard practice in the region.

Specifications

The specifications are a second part of the construction documents. They can be subdivided into two types: performance specifications and materials specifications. Performance specifications describe and control the work performance itself, while materials specifications control the quality of the materials used. For example, a performance specification for concrete might specify the conditions (including moisture and temperature) under which the concrete is to be poured and the particular curing method to be used. The materials specifications for concrete would specify its required strength (in pounds per square inch), its stress resistance, its slump characteristics, its compositional mix, and so forth.

A uniform system of specification (which is in fact called the Uniform System for Construction Specifications) has recently come into use through the joint efforts of the American Institute of Architects (AIA) and the Construction Specifications Institute (CSI). This system presents a consistent order of subdivisions for each of the sixteen categories it specifies.[3]

Specifications take precedence over drawings in all cases. This means that, if there is a discrepancy between a material or method of work as drawn and as specified, the contractor should follow the specifications rather than the drawings.

Contracts

The actual contracts are also a part of the construction documents. The contracts include the agreement, the contractors' requirements for posting performance and labor and materials bonds, liability and insurance clauses, general conditions pertaining to a variety of matters (the notice to proceed with the work, time to completion, bonus or penalty clauses, the fee schedule, temporary services during construction, and protection of adjacent facilities), change orders and other amendments to the contract, and guarantee provisions.

The bidding documents are often a part of the "front-end" information in the contract document set. These include the invitation to bid and instructions to bidders, the bid bond forms (if a bid bond is required), and the bid form itself. The instructions to bidders address the manner in which bids are to be received, including a clause reserving to the owner the discretionary power to reject all proffered bids.[4]

Once construction documents have been completed, they are

usually sealed by a registered architect, landscape architect, or engineer (depending on the project and the requirements of the job) and then issued for bidding or negotiation, which is the next stage of the design process.

BIDDING OR NEGOTIATION

Negotiation is sometimes considered simpler than bidding as a means of procuring a contractor for site work—although this is not always true. A job may be negotiated for a number of reasons: the owner may have had previous experience on a project with a particular contractor and want the same contractor to do the additional work (this is common on phased projects occurring over a period of time); or the choice may hinge on real or perceived favors owed by one party to another; or negotiation may be suggested by expediency, so that the job will not be delayed for the three to four weeks required to go through the bidding process.

Negotiation is essentially a bartering process in which the owner and the contractor submit, reject, and modify offers until an acceptable compromise is reached. This does not necessarily mean that the owner will be getting the most work for the least money, or (for that matter) the best work. Nevertheless, for the above-mentioned reasons, many projects are built on a negotiated basis.

Bidding is generally more involved and time-consuming than negotiations. In the first place, in order to be as fair as possible to prospective bidders, the construction documents must be concise and complete, whereas the same need for meticulous detail may not exist in a negotiated project. The bids are usually advertised for a specified amount of time, which varies depending on whether the job is governmental or private. The time, place, and method of receiving bids, as well as the decision to conduct the bid opening in public (or not), are given in the instructions to bidders.

Once bids are received, they are compared to ensure that each contractor bid on the same thing; the (apparent) lowest and second lowest bids are then noted—that is, recorded in written form. It is customary to take the lowest bid, if it is a good bid. Determining whether a bid is good or not involves such considerations as the accuracy and completeness of the bid, its offering an acceptable base price (the lowest bid may still be too high),

and its fulfilling certain technicalities of form. The award of bid often requires some additional negotiating, either on price or because of substitution of materials. The award of bid is followed by the signing of contracts and the notice to proceed with construction of the project.

CONSTRUCTION

Once construction (the next stage of the process) begins, changing the design becomes exceedingly difficult and costly. The contract will stipulate just what approval is needed before necessary changes can be made. Usually the owner defers to the designer in terms of the desirability of a change, but the owner's written approval is required for any monetary changes. Since the designer and contractor both work for the owner (and not for each other), a change requested by one of them to the other is not legal without the intervention of the owner. Necessary changes are legitimated through a change order.

The designer may be entrusted with duties during construction ranging from continuous on-site inspection to inspection only at critical periods. The final inspection of the project is accompanied by a punch list—a list of all incomplete or unsatisfactory items that need fixing before the project is finished and before final payment will be made.

Postconstruction Services

Construction is often followed by postconstruction services, one of which is enforcement of the guarantee provisions. This can often take years. In the case of vegetation, the guarantee is usually for one year for full replacement, and two years for partial replacement. Other postconstruction services include evaluation of materials or methods of construction (often done by the designer for his or her own future reference, without being paid), life-cycle costing, energy studies, and expansion or future design work.

The designer can become involved in a project at any of the stages described above; and since the process is cyclical, it allows for feedback and revision along the way. The more involvement allowed to the designer from the start, the more likely it is that the project will evolve as a sensitive accommodation of site, space, and structure (figure 1-14).

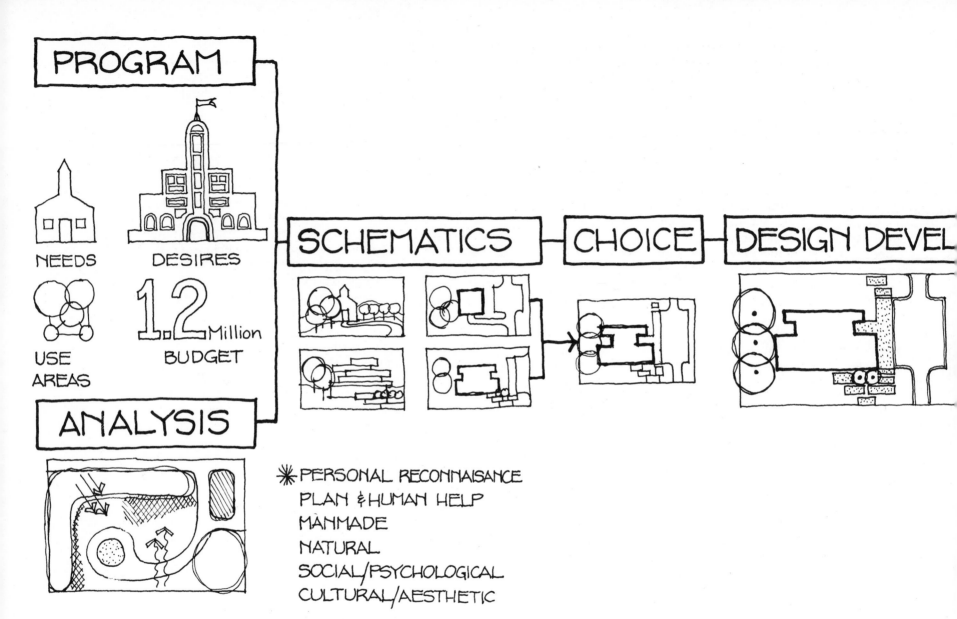

PROGRAM

NEEDS

DESIRES

USE
AREAS

1.2 Million
BUDGET

ANALYSIS

SCHEMATICS

CHOICE

DESIGN DEVEL

✳ PERSONAL RECONNAISANCE
PLAN & HUMAN HELP
MANMADE
NATURAL
SOCIAL/PSYCHOLOGICAL
CULTURAL/AESTHETIC

1-14. A summary of the design process.

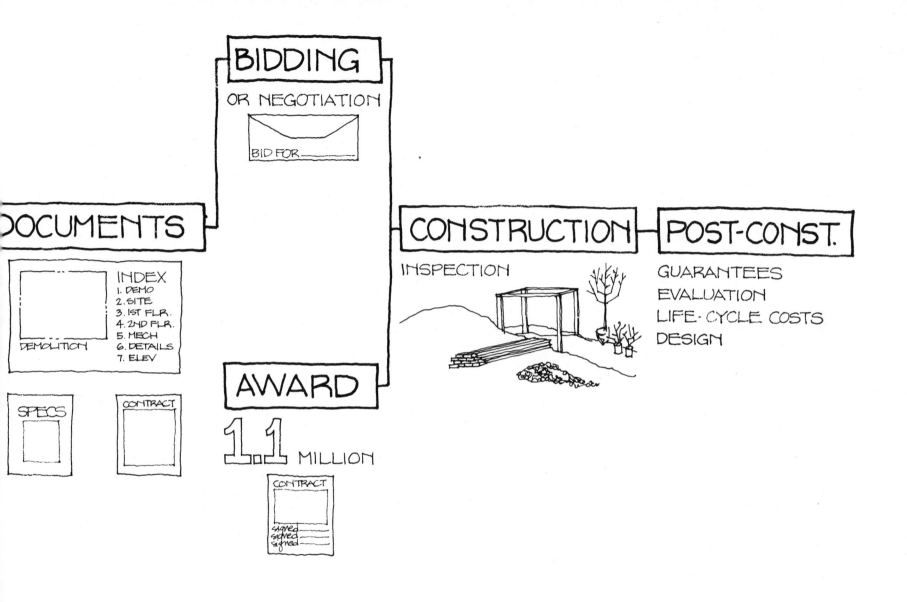

BIDDING

OR NEGOTIATION

BID FOR

DOCUMENTS

DEMOLITION

INDEX
1. DEMO
2. SITE
3. 1ST FLR.
4. 2ND FLR.
5. MECH
6. DETAILS
7. ELEV

SPECS

CONTRACT

AWARD

1.1 MILLION

CONTRACT
signed
signed
signed

CONSTRUCTION

INSPECTION

POST-CONST.

GUARANTEES
EVALUATION
LIFE-CYCLE COSTS
DESIGN

REFERENCES

1. Garrett Eckbo, *Urban Landscape Design* (New York: McGraw-Hill, 1964).
2. Lawrence Halprin, *The RSVP Cycles: Creative Processes in the Human Environment* (New York: George Braziller, 1969).
3. Harlow C. Landphair and Fred Klatt, Jr., *Landscape Architecture Construction* (New York: Elsevier North Holland, 1979).
4. Landphair and Klatt, *Landscape Architecture Construction.*

2 | Space

The process discussed in the preceding chapter allows a designer to develop a site design that integrates site, space, and structure in a functional, interesting, and aesthetically pleasing way. A prerequisite for such use, however, is a clear understanding of the concept of space.

The notion of three-dimensional form is fundamentally dependent upon the concept of space, which is first of all defined by means of lines and line associations (a line being the joining of two points). Lines themselves have certain shapes, and these shapes are capable of imparting certain feelings or qualities to the observer. A line can be flowing and smooth, as are the lines of rolling hills disappearing in the distance, or it can be sharp and angular in the manner of jagged mountain peaks. It can be tranquil and relaxing, so that it encourages quiet, calm times, or it can be fraught with conflict, causing feelings of uneasiness and unresolved problems (figure 2-1).[1]

The quality of the line determines in part the qualities of the forms, because forms are created out of lines. A smooth, organic form cannot be created by using an angular line, nor will a firmly disciplined form be produced out of a weak, curvilinear line.

The form that is created in two dimensions is the plan representation of the three-dimensional reality, seen first as a mental image in the mind of the designer and later as a finished product. If the lines are insecure, uneasy, and unresolved, the forms resulting from them will be the same way, and the conflict in the design in two dimensions will not be resolvable in three. If anything, a plan-form relationship that seems to work on paper becomes more difficult to make work in three dimensions, not easier.

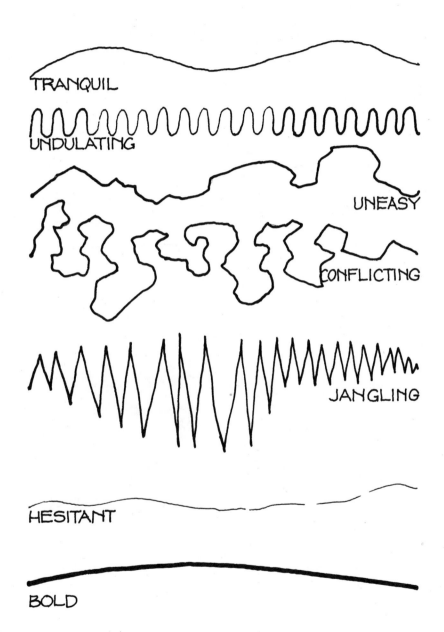

2-1. Lines represent moods.

Since the two-dimensional forms created by certain types of lines represent three-dimensional spaces, the designer cannot expect to create a pleasing composition by thinking in two dimensions: the thinking must include volume from the start, and some sort of capacity for containment must be added to a two-dimensional object defined by lines before it becomes a volume. A flat piece of cardboard can contain nothing, but when sides are added to the cardboard, enclosure results.

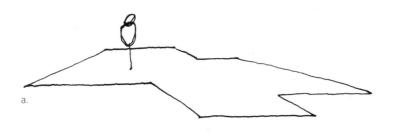

a.

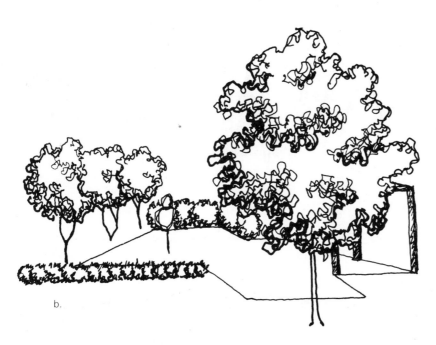

b.

2-2. Unplanned space compared to planned space: a. negative, unenclosed, unplanned space; b. positive, enclosed, planned space.

NEGATIVE AND POSITIVE

Traditional design considers the physical object to be a positive element and space—the blank or unfilled area around the object—a negative. In site planning, space becomes a positive entity, too, supporting and enhancing the appearance and function of the objects or structures within it (figure 2-2). However, negative space does occur in site planning. Any space that is unplanned, unenclosed, or unintended for human use constitutes negative space. An example of negative space is provided by an architectural structure placed dead center on a site, usually on a mound to elevate it above its neighbors. The space surrounding that building—since it relates not to the structure, but to the outside elements—has been wasted (figure 2-3). The same design concept could be used successfully to create positive space, however, by relating the height of the structure to the heights of surrounding elements, and by relating the open space and detailing to features in the surroundings.[2]

2-3. Negative space.

The concept of space as a potentially positive entity takes some mental adjusting to comprehend. Contour drawing, in which the pencil follows the outline of a series of objects—in effect defining the objects by delineating the spaces not filled by them—is one way to begin to understand space (figure 2-4).

FORM RELATIONSHIPS

Since space is experienced not as a composite of individual frozen sectors but as a whole, the relationships between forms determine the character of the space as a whole. The forms in a spatial design can be similar in size, shape, or placement, or very dissimilar. As forms are placed beside and over and under one another, a series of layers of space and form are created. This spatial stratification makes the experience of space meaningful, setting the user behind, under, above, or against certain spatial patterns (figure 2-5).

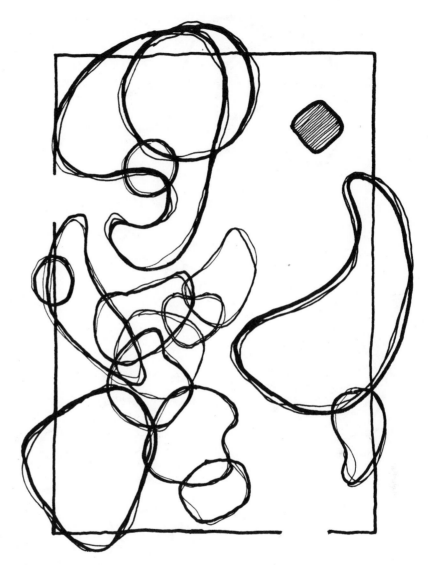

2-5. Creating different spatial experiences by stratifying elements.

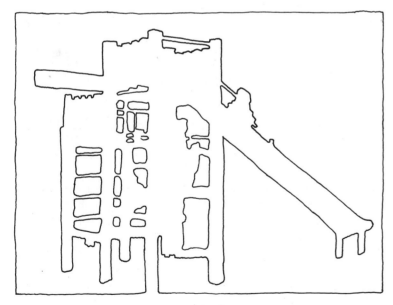

2-4. A contour drawing, delineating *object* by *space*.

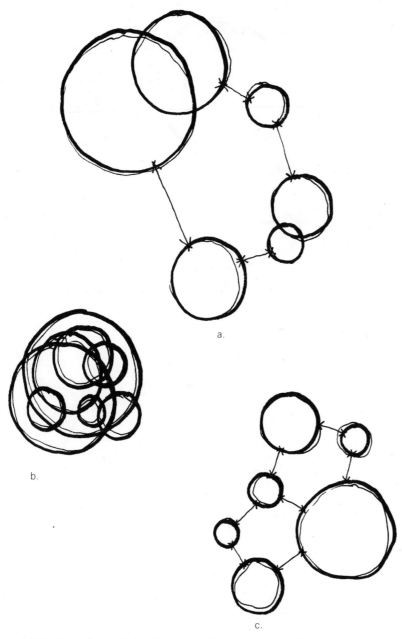

2-6. Space-form relationships: a. sensitive placement of forms; b. equidistant forms floating in space; c. densely stacked forms creating confusion.

Forms, as volumes, must be considered in terms of their total size and in terms of the relationship of a form of one size to a form of another. Consideration of form relationships must take into account the spaces between the forms, which are themselves forms of another type. The extremes of form placement vary from spacing each element equidistant from the next, to making them float like dumplings in soup, to stacking them on top of one another like poker chips (figure 2-6).

The two-dimensional representation of plan form does not indicate the densities, textures, and degrees of actual enclosure that occur in three dimensions. A balance must be achieved between light and dark, among colors and textures, and in the scale of the forms. The use of all of these design characteristics determines the primary focus of the site and spaces. There may be a primary focal point or several smaller ones. If several exist, the user should still be given a clear choice of experience. Spaces created by line and form relationships that are sensitive to all critical design characteristics give the users a choice of direction and visual experience, of openness or enclosure.

Too many designers skip the stage of spatial design that really involves schematic design (the stage described and shown above) and let a detail dictate the entire site design. Even though detail is important in creating unity of experience, a site plan that evolves from the desire to use a certain detail may fail.

A typical design example of this is a small park, with two entrances, intended to be used for a variety of self-guided activities such as Frisbee or football, picnicking, reading, and walking. Seeing the two entrance points, some designers immediately place a path between those points and then, using the lines that define the edges of the path, begin placing the detail elements such as trees and shrubs, picnic spaces, and fountains. These designers have not designed space based on form relationships; instead they have started with two discrete lines and gone to innumerable details without articulating the form the space is to take and without deciding whether the human element or the site is to dominate. As a result, the path effectively dominates (figure 2-7).

Just as buildings designed by different architects to serve different functions have their own unique, individual characters, so different spaces have their own special design form and qualities. Space takes its form from the objects that surround it and create it—the buildings, the sky, the trees. Space both allows the existence of objects and is dependent for its own qualities on those objects. This interdependence between space and object is more important than the qualities of either taken separately, because the separate qualities are not perceived by the user.[3]

PERCEPTION AND REALITY

A person's perception of space depends on his or her size, age, cultural upbringing, mood, past experiences, and expectations. To a small child, distances are vast, heights are unreachable, and separate worlds exist under every shrub and around every corner. To an elderly person, distances may be lengthened by weather conditions or slope, and enclosed space may seem more frightening than reassuring.

The American notion of the front yard as a nonfunctional showcase is totally alien to some other cultures, where every inch of land is put to productive use. A person from a rural background might look at spaces from an economic or conservationist standpoint, whereas a person used to an urban setting might see only their development potential. A winding country road lined with pleasant fields and thickets is a delight to a person in a relaxed mood; to someone in a hurry, though, it is an exasperating obstacle.

Past Experiences and Future Expectations

The past experiences and future expectations a person brings to a site significantly influence the person's use and perception of the space. People relate to present situations by past experiences and future expectations. For example, a person from a small town making an annual visit to a big city for a shopping spree may find everything about the trip exciting because the city seems to possess a comparative wealth of size, colorfulness, and quality. On the other hand, a daily commuter to the same city experiences the trip as a time-consuming hassle—boring but necessary.

Another example might be a visit to a renowned five-star restaurant. The diners anticipate an experience of culture and qual-

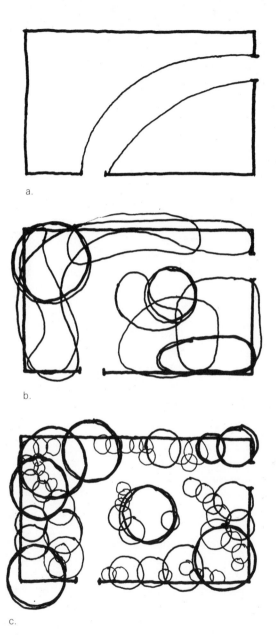

a.

b.

c.

2-7. Comparing space defined by a detail and space designed according to form relationships: a. the lines of the path dictate the design; b. a form study as the schematic basis for the design; c. spatial character based on the schematic form study.

ity, but if their drive takes them through a run-down industrial section of town (where the restaurant is located), doubts may arise as to the quality of the fare they should expect to find at journey's end.[4]

Design Effects on Perception

The designer's analysis of the conditioned response of a person to a site should begin with a careful assessment of where people come from to reach the site and where they go when they leave it. What people see, hear, and think along the way affects their feelings toward the site itself; people cannot help comparing one place or experience with others.

Sensory Impressions

People also entertain concepts of space that are based on sensory impressions at a given time. A place may be inviting because the scent of lilacs or freshly baked bread draws the person to it, or it may be repulsive because of the repellent smell of sewage or corn processing. The physical design of the space may be exactly the same in each case; it is the sensual perception of the space's quality that encourages approach or avoidance. Likewise, noise from flight patterns or expressways may diminish the value of a site (regardless of how efficient, safe, and visually interesting it otherwise is) because the site's actual material character is subordinate to its perceived discomfort.

The Feel of a Site

Although most people are so dependent on their eyes for orienting themselves in space that they pay little attention to what their other senses tell them, even a fully sighted person recognizes in a site a certain feel. Without using the eyes, the body can sense its proximity to other objects or people—to intense light or dense shade, to threatening overhangs or abrupt cliffs, to cold and hot.

A designer can capitalize on this sensory experience of space through subtle manipulation. Hot spaces can be made to seem cool through the addition of moving water (with its sound, smell, appearance, and actual cooling effect) or through the use of masses of dark plant materials. A skyscraper environment can be humanized by introducing elements overhead and to the sides that make the body feel that it is in scale with its surroundings—protected by what is beside it and by what is overhead.

SPATIAL MODIFIERS

Space is modified by practically everything: the quality of the spaces and objects that surround it; the colors, height, breadth, and degree of maintenance of nearby buildings; climate and seasonal change; light and dark; the routes by which people enter and exit the site; the presence or absence of people or other moving, living things on the site. Successful spaces are designed to adapt to the changeable conditions around and within them, even though the designer may have little or no actual control over the modifiers themselves.

Climate and Seasonal Change

The modification of spaces by climate and seasonal change can be extremely dramatic. A blinding snowstorm reduces a familiar location to an incomprehensible and dangerous wilderness, although the first fat lazy flakes of the season send child and parent alike running delightedly for the sleds. The absence of wind from an intimate sheltered garden can turn the space into a sweltering pen, but conversely the unobstructed sweep of wind across an exposed plain can psychologically increase the distance between points of shelter.

Quality

The quality of the surrounding spaces and objects has an impact on the space of the site itself, since space takes its character from its surroundings. If the quality of the environment is uniform, cohesive, and quiet, the space of the site will assume those same qualities. A run-down, uncared-for environment may give the site a shabby appearance, regardless of the care lavished on it, or (viewed the other way) the site may stand out in such sharp contrast to its surroundings that they look shabbier still.

The colors of a site also affect the feeling of quality it conveys. Color modifies a site by creating unity and harmony or contrast and clash. The colors not only of the structures around the site, but of all other nearby elements—pavements, signs, plants, vehicles—influence the perception of the colors used on the site itself. The colors of all elements on the site may have been chosen with the utmost care and with an eye toward creating a pleasant composition, but if they clash with a background element or are overwhelmed by something of a different color, the design effort is wasted.

Surrounding Structures and Elements

The height, breadth, and overall character of surrounding structures and elements modify space by lending to it different degrees of density or openness, by appearing imposing and threatening or harmless and sheltering, by acting as a neutral backdrop for activity in the space or by being so interesting or incongruous that anything occurring in the space seems subordinate to the background. Height and breadth contribute significantly to the feeling of scale that articulates the character of the space (figure 2-8).

Light

Spaces are modified in a number of ways by the quantity and quality of light reaching them. Light may be direct and glaring (causing users to shield their eyes and seek shelter in the shade), or it may take the form of shadowless north light (offering users no visual clues as to the time of day), or it may consist of "bounce" light reflected from adjacent structures or pavements and controlled by on-site design considerations (creating for users a glare-free, pleasant, bright atmosphere).

The quality and quantity of light change as days lengthen in the springtime, and even as the sun moves across the sky from morning to evening. The character of a space thus changes over the day and over the year. A path that is safe and acceptable during the daylight hours can take on sinister qualities when negotiated in total darkness; a place of garish colors and textures can become enticing and infinitely interesting when softened by low-level lighting; and a road that is harsh and forbidding under sodium lights can completely change character when lighted incandescently.

Approaches

Spaces change in appearance depending on the ways by which people get into and out of them, and depending on the vantage points from which they are seen. A space approached from below (up a series of stairs from which the space itself cannot be seen) will leave a different impression on the user than the same space approached from eye level. A person with a

2-8. The character of the surroundings modifies space: a neutral structure as the background of a busy space.

garden apartment experiences space literally at ground level, which means that the vertical dimension of the space between apartment buildings becomes exaggerated. A person with a twentieth-floor penthouse is divorced from any immediate sense of the land, but the boundaries of the site extend far beyond the legal limits of the property.

The main approach to a site can be grandiose, formal, and direct, focusing on an imposing terminating element, or it can be a winding path offering ever-expanding glimpses of what lies beyond the last bend, filling visitors with a sense of increasing anticipation. The designer may want the actual entrance gates or doors visible from as great a distance as possible, so that the location of the final destination is never in question, or the aim may be for visitors to come upon the entrance suddenly and as if by surprise. The approaches and viewing points may be calculated to make people feel awed and insignificant or to make them feel immediately at home.

The mood of a space is often established long before a person physically enters the site. How the site fronts on the street or walk, whether its entrances are large and open (inviting use) or are small and solid (suggesting private ownership), and whether anything of the character of the site is revealed to the casual passerby all help to determine the ways in which people will use the site.

Progressive realization can be used in conjunction with an understanding of conditioned perception in designing the mood of the site. Progressive realization is the viewer's response to the gradual revealing of a site or site feature, so that the person approaching becomes increasingly aware of the detail rather than the entire design. An example of this might be experienced in approaching a monument that, from a distance, is visible in its entirety against a background of other buildings and spaces (even though these, if seen in context, might actually dominate the scene). As the person moves closer, the monument becomes progressively difficult to view as a whole, until finally the stonework of the entrance is the only thing in the person's immediate field of vision, and the original background has become invisible (figure 2-9).[5]

2-9. Progressive realization changes the appearance of a site feature.

People and Life

Spaces are changed instantly by the presence of people, animals, or other moving, living things. The lone user of a large and open site may feel uncomfortable, as though to occupy that space were to do something forbidden. Populate the same space with laughing, eating, drinking, talking, moving human beings, and the scale, mood, interest, and use of the space all change.

Many exterior spaces seem as though they were really designed to be just as blank and lifeless as they appear in architectural photographs (from which people are often excluded because they add too much disorder and variety), yet people will attempt to occupy even such inhospitable barrens of concrete and glass, if they are given no alternative.

The basic premise of site planning—and especially of site planning that integrates site, space, and structure into a unified whole—is that spaces are to be designed for human occupation. A modest-size space can become woefully overcrowded when too many people try to use it at once. A child's playground is a piece of interesting but useless sculpture until it swarms with children playing. A space designed for people to use seems incomplete without them.

WHY DESIGN SPACE?
Comfort and Security

The overriding reason for designing either interior or exterior space is to give the people using it a sense of comfort and security. People, like other animals, derive security from their relationship to their environment. Anything in that environment that creates in people a feeling of uneasiness or discomfort will detract from their ability to use the site in the ways for which it was intended. An intruding element, an incomprehensible scale, or too much clutter may lessen the efficiency or productivity of the site, causing people to seek satisfaction of their needs elsewhere.

Creating Choices

Any design for human use should certainly allow the functions of the site to be performed. Beyond that, however, the design should offer choices to the users without creating chaos. Arrayed with two or three logical, perceptible choices of directions to take, areas to use, or things to look at, a site becomes interest-

2-10. Pleasant choice (top) versus confusion (bottom).

ing and inviting. Too much of a good thing, though, causes the number of choices to become overwhelming, and what was intended as a variety of choice becomes no choice at all (figure 2-10).

Scale Relationships

People gain a sense of comfort and security from their environment if their physical relationship to it is in scale according to their own perceptions. A space that is wide open instills in viewers a sense of their own smallness, or perhaps a sense of awe. A small space creates a feeling of human dominance, intimacy, or crowding. Large spaces can be made to feel smaller, and small spaces expanded, by varying the degree of containment or enclosure used in creating them. The key element in establishing scale in the design of space is enclosure.

ENCLOSURE
The Elements of Enclosure

The elements by which enclosure (whether for exterior or interior space) is established are the base or ground plane, the overhead or ceiling plane, and the vertical or wall plane. By using different combinations of materials and different degrees of dominance in these three planes, the designer varies the quality and function of spaces.

The Ground Plane

The ground plane is the most important plane in terms of function. Inadequate accommodation in the ground plane for the intended use of the space cannot be rehabilitated by even the best design in the other planes. An example of this is circulation patterns in a public park. If the sidewalks are not located where the use patterns show them to be needed, the walks will go unused and the dirt paths will remain. It is true that paths can be blocked with vertical walls or solid barriers, forcing people to alter their natural routes and use the paved surfaces, but this should point out to the designer an evident defect in the design: the ground plane, on which people travel, was originally designed without sufficient regard for function (figure 2-11).

The quality of experience established by the lines and forms in the ground plane suggests the quality of experience to be expected in the rest of the space. If the lines of the ground plane are subtle, smooth, and flowing, blending imperceptibly with one

2-11. A dysfunctional ground plane is apparent by the well-developed path.

another, people will move through the spaces smoothly and quietly. If instead the ground plane presents a series of sharp angles, each of which attracts the users' attention for a fraction of a second and forces a decision, the space will take on a much more conflicted form (assuming the design in the other planes complements the design in the ground plane).

The materials used in constructing the ground plane should be natural or should look as though they come from the earth and belong naturally on or in the ground. This is in keeping with cultural and social expectations: people are accustomed to walking outdoors on surfaces that are rough or dull, brown or gray or green, but not on surfaces that are mirrored or glass-smooth or bright yellow or purple.

The Overhead Plane

Although in the design of interior space the overhead plane provides important protection from the elements, when used in exterior space it should be subtle enough to allow the user merely to sense what is overhead rather than to become consciously aware of it. This can be accomplished by paying careful attention to the density of the overhead plane and to the amount of open space between the overhead plane and the observer. If the overhead material is low, solid, and dark, its apparent mass will increase and the apparent open space around the user will decrease, making the space feel as though something were pressing down on it. If the overhead plane is too far removed from the viewer, the structure's scale will be perceived in relation to closer verticals, thus negating the need for the overhead as a spatial definer (figure 2-12).

An overhead canopy in exterior space serves the valuable function of drawing the vertical scale of many man-made elements such as skyscrapers down to a human level. Likewise, it offers much-needed psychological (if not actual physical) protection when it is used as a definer of space. For example, pedestrians feel more secure on a sidewalk that directly abuts a road if they feel shielded beneath the overhang of a tree than if they are exposed to the sky—even though the actual distance between the walk and the curb is the same in each case (figure 2-13).

An exterior overhead plane can also offer a fair measure of protection from the elements, extending the protected area's season of use, and can carry the architecture outward into the site, thereby integrating the two more fully. As simple a design feature as an extension to a roof overhang can provide a place for two or three people to stand and watch the rain on a wet, warm day or a place for them to wait in cold weather. Gradually decreasing the solidity of the overhead plane can provide a transition from solid roof to open sky and can create interesting experiences for those underneath. The suggestion of a canopy— through use of an eggcrate lattice, an arbor covered with vines, or a sheltering, spreading tree—encourages people to move through the intervening spaces to reach the shelter. People associate pleasant spaces with comfort and safety, and a roof over one's head has long been a measure of both.

Finally, the overhead plane affects the quality and quantity of light that reaches the spaces beneath it, casting shadows, changing patterns depending on sunlight or cloud cover, emphasizing the ground pattern or imposing a new texture on it. Anything that affects the light reaching a space also affects the character and mood of that space. The shadows may be strong

2-12. A successful overhead plane—suggesting enclosure but not crushing the area under it.

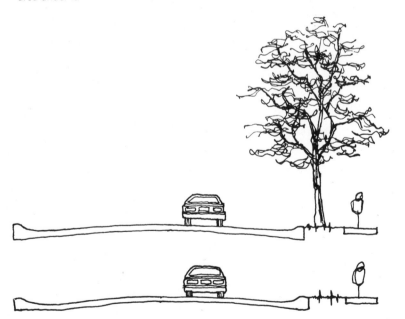

2-13. Psychological protection provided by an overhead canopy.

enough elements in themselves to cause use patterns to develop in association with the areas in shadow and the areas in sun.

The striking effect of an overhead canopy and its shadow is immediately apparent on a hot summer day when a pedestrian passes a heavy arch and becomes aware of the cool darkness of the alcove behind it—a coolness implied by the shadow from the arch. The pattern of the overhead plane reproduces itself on the ground plane, either emphasizing or blurring its definition. The fine tracery of leaves may be the only pattern apparent on a blank pavement, or a delicate latticework structure may mimic the joints in a modular brick paving. A flowing, curved line in the ground plane might be copied by the line of a support for a trellis, giving the user a more evident feeling of movement along a desired path.

The overhead plane can also overpower the ground plane if the shadows cast by it are so heavy that a person's attention is attracted first downward toward the preponderance of dark and light, then upward to see what is causing such a heavy texture. The heavier the overhead plane appears, and the greater the spacing between members, the greater the shadow pattern's contrast and dominance will become. When this effect occurs, the apparent vertical distance between the overhead and base planes appears drastically reduced, sandwiching the space into a much smaller dimension. Needless to say, such a heavy hand with shadow texture is to be avoided on small sites, and used with caution on larger ones.

The most beautiful overhead planes are natural ones, over which the designer has no control other than to allow them to be seen. These are the planes of the sky at sunrise or sunset, the incredible blue of an Indian summer day, the breathtaking sight of thunderheads building in the distance, the shimmer of stars in a night sky so black its texture can nearly be felt. The designer's control of the site-space experience should not be so tight that these experiences are denied.

The Wall Plane

The vertical or wall plane is the plane over which the designer has the most control and by means of which the greatest degree of enclosure can be achieved. Because verticals are used in both interior and exterior spaces to create enclosure, both types of spaces take much of their character from the verticals. Vertical enclosure can be subtle, rising from the ground plane or extending from the overhead plane in such a way that the place where one plane begins and the other ends is not immediately visible, or it can be bold and obvious, as in a free-standing vertical wall or a series of columns or narrow upright trees.

Placement of the vertical elements in space should defer to the function needed in the ground plane. For example, the designer might incorporate verticals along the outer edge of a curved roadway to direct the attention of the driver around the curve rather than out across the landscape. Verticals may accentuate an already dominant ground plane feature, such as a plateau or knoll, or they may be located to give visual relief to a flat, monotonous site. In any case, the placement of the verticals should be handled to enclose the site, enhance the structure, and create a smooth, logical flow between the spaces.

Verticals can serve as a backdrop for important specimen features (in which case the use of neutral materials, soft colors, and smooth transition lines is important), or they can serve as focal features themselves—directing the viewer's attention away from a questionable area. They can screen or block views and can help control climate. Because the scale of most landscapes is so distinctly horizontal, vertical elements take on exaggerated importance and appearance, especially when they occur in isolation. A specimen tree silhouetted against a gently rolling prairie, a single sentinel row of telephone poles, and the spire of a building all act as focuses and reference points in their respective landscapes. The designer turns to the vertical plane to create points of reference or landmarks, and people use these features to orient themselves in space.[5]

THE SCALE OF EXTERIOR SPACE

Although the same three planes articulate both interior and exterior spaces, the scale at which the design is accomplished is significantly different for the two types of spaces because the point of reference for the user is so different. In interior space, people assimilate the environment to their own heights, to the heights of vertical walls, and to standard door and window dimensions. In exterior space, the vertical reference elements may be skyscrapers rising over two hundred feet, a distant mountain range, or a fifty-foot-tall tree, and the horizon line may be miles away.

Exaggeration in the scale of exterior space is therefore needed to make a person feel as though the space is sufficiently large to accommodate comfortably uses corresponding to similar indoor activities (figure 2-14). A rule of thumb is that eight to ten times a given indoor area is needed outdoors to provide the same feeling of room for a particular activity. For example, an indoor dining area that seats eight (about ten feet by twelve feet) would need to be twenty-five feet by forty feet to provide the same feeling of comfort for eight people outdoors.[7]

2-14. Comparing interior and exterior scale.

ALTERING THE ENCLOSURE

The sensitive use of varying degrees of enclosure—from almost complete (complete enclosure does not really occur in exterior space because complete enclosure implies four solid walls and a solid roof) to almost nonexistent—allows very different spaces to be developed on the same site. Since people have no immediate perception of spaces that they do not occupy and cannot see (in other words, spaces on the other side of a definite boundary), design can differ radically in different site areas, with each part of the site designed to take best advantage of program requirements and site conditions.

Rarely will almost complete enclosure be desirable because a site totally walled in by enclosing elements turns its back on the outside world. By providing openings in the verticals so that users' eyes are directed toward interesting off-site features, the site can be made to extend beyond its legal boundaries, and the reach of experience can be made to surpass what might otherwise be available.

Distance-Height Ratio

The key to successfully altering the degree of enclosure lies in the distance-to-height ratio between enclosing elements. A ratio of one to one—the enclosure height being equal to the open space distance—is a critical figure: when the height exceeds the open space distance, the enclosing elements become overpowering (with the result that users are more aware of them than of the space itself), and the verticals take on exaggerated importance as overheads. This type of space can make users feel as though they are in a canyon or at the bottom of a well, thoroughly subordinated to their surroundings. Such spaces can seem richly human, however, depending on how the human-scale elements of the space are handled.

If diversity and variety are supplied at eye level, and for a short distance above and below eye level, attention will focus on that zone, and the bulk of the verticals rising above it will be negated. A space of the same scale treated by turning blank, uniformly monotonous walls toward the users, however, will appear forbidding and dark, and people will use the space only as necessary to get somewhere else.

Another critical figure occurs at a ratio of four (open space

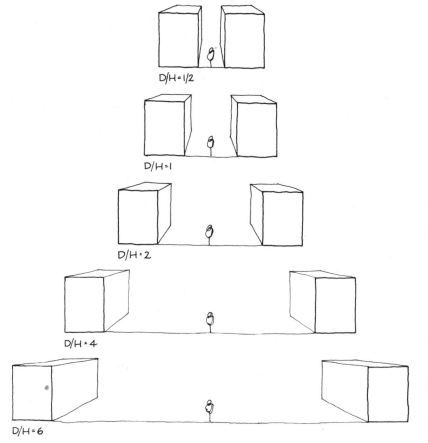

2-15. Spatial scale as determined by distance-height ratios.

distance) to one (enclosure height). Above this ratio, the degree of enclosure has diminished to a point at which it ceases to exist and the user must seek elsewhere for an element by means of which to establish spatial scale (figure 2-15).

The most comfortable, workable exterior spaces are those in which the ratio of the open space distance between vertical elements to the height of the vertical elements is between one and four to one. The best-designed, most dynamic plazas and open spaces in the world have distance-height ratios of two or three to one.

The same ratio relationship works in regard to openings in enclosing elements such as walls, hedges, and plant material masses. If the opening is significantly wider than it is high, the effect of a gate or passageway will be lost. A very narrow opening, with the ratio of distance to height that is below one to one, will downplay the importance of the opening, inviting only those who belong to enter (figure 2-16).[8]

Tricks

If for some reason exterior spaces cannot be designed to an acceptable distance-height ratio, tricks can be played on the user's perceptions of those spaces to increase or decrease the apparent height of the enclosing elements, as desired. Using enclosing elements that have vertical lines themselves (whether they be the form lines of a concrete wall, the slats of a fence, or the columnar growth habit of certain varieties of trees) tends to emphasize their height, leading the eye upward more dramati-

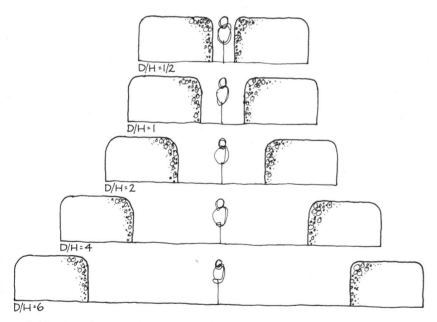

2-16. The relationship of distance-height ratio to openings in the enclosure.

2-17. What appears to be a long space receding in the distance is a very short space made longer by converging rather than parallel side lines.

cally. The use of horizontal lines in stone coursing, of even, rounded hedges, or of horizontally placed railroad ties or steps causes the eye to relate the vertical elements to the ground plane rather than to the sky. The length of a short space can be extended visually by using fine-textured materials in the background and by imperceptibly changing the angle of parallel lines in the ground plane—causing them to converge at some extended distance rather than to remain truly parallel (figure 2-17).

Accuracy in Exterior Space

Directly related to the expanded scale of exterior landscapes is the designer's increased leeway in deviating from exact accuracy. Since the scale and frame of visual and physical references are larger, control of detail can in most cases be looser than would be possible indoors. A gradual wiggle in an otherwise straight sidewalk running for four or five blocks is likely to be indiscernible to anyone but the designer; the same wiggle scaled down to fit in a building corridor, however, becomes totally unacceptable.

The designer's freedom to deviate rapidly diminishes as the user experiences the exterior space on a personal scale. For example, a person naturally slows down and nearly stops at the main entrance to a building, in order to open the door. In such a location, a flaw in the alignment of the paving pattern or an awkward bench detail or construction joint will be immediately visible.

THE VISUAL QUALITIES OF SPACE
Dominance

Much of a person's experience of space is visual. In order to capitalize on the appearance of site, space, and structure, the designer must first determine whether the character of the spatial elements or of the human activities in the space is to dominate. This decision orders the scale of the spaces designed and of the details used, selects the locations and appearance of approaches and exits, and determines the overall quality of the space.

If the space is to dominate, the design effort should be directed toward creating feelings of humbleness, smallness, or subordination. All of the elements of such a site should be designed with qualities that give them preeminence over any activity that might go on in the space. Many natural phenomena are site-dominated rather than people-dominated—the Grand Canyon, Niagara Falls, Muir Woods, the Badlands. Likewise, the cityscapes of major cities dominate and subdue: imagine the feelings of immigrants seeing New York City from a distance for the first time.

People-dominated sites should be designed to accommodate the activities intended for the site and structure, whether in a passive or vibrantly active way. The colors, scale, enclosure, transitions between open and enclosed spaces, and materials should all be chosen to make the people using the space comfortable and should encourage and support the activities rather than conflict with them. Such spaces may be more neutral, passive, and self-effacing than site-dominated spaces, taking their character from the ever-changing kaleidoscope of human activity.

Many spaces that are intended to be people-dominated fail because of built-in design conflicts, such as children's playgrounds paved with concrete, surrounded by chain link, and overlooked by blank, impersonal windows, or parks intended as recreation areas but posted with "Keep Off the Grass" signs and edged with barriers that deny free access to the inviting grass. If a people-dominated space is so blank as to be devoid of interest, or is too riddled with cautions, or exhibits a scale, color, or design character that suggests some other use, people will not take possession of it and dominate it as intended.

The views of, to, through, and out of spaces play an important part in determining whether a setting is site- or people-dominated. Put people in the bottom of a fishbowl of office complexes, and then design the space to encourage privacy, and no one will use it. Allow the view to the site to change constantly as the user moves through it (keeping the final destination ever-hidden but beckoning), and the whole spatial experience will be rich with anticipation. The sequential observation of a site not only builds a sense of anticipation toward what lies ahead, but creates for the user a feeling of orderly progression. If each sequential view grows logically out of a preceding view and space and suggests the view to come, the space will be experienced as a single coherent entity.

Designers seem often to forget in the two-dimensional design of space that, of all design characteristics to be considered in three dimensions, the visual quality of the space is most important in terms of creating feelings of scale and comfort. They may design a magnificent, breathtaking view of the major site feature—from only one direction, ignoring the fact that people approach a site from many directions at many different speeds, and that all possible views to and from and without and within need to be considered. There is nothing more disappointing than entering a site from the "proper" direction one time (and finding the experience pleasant and rewarding), and the next time trying a different approach only to discover that the view from this direction was obviously not designed but just happened.

View and Vista

Space can be seen in two primary ways: by means of a view or by means of a vista. A view is a panorama from a given point, sweeping across a considerable area and not really focusing on any one site feature. The view of a site and the site's use must be compatible if they are to give the user the richest possible experience. A view also changes as a person moves through the spaces, making the total sequence of views more important in determining the character of the space than any single vignette.

A vista is a confined view, usually directed toward a terminal space or element. It is controlled in its entirety by the designer. A common mistake in attempting to create a vista is forgetting that the space is likely to be experienced from more than one direction. If the direction of movement is toward the terminus, the vista from the terminus back along the path must also be controlled.

A vista may be part of an overall view, experienced perhaps on a smaller scale. This may be accomplished by framing a vista within a view, using enclosure and directed attention. A vista need not be imposing, impressive, or site-dominated: it can be designed on a very small scale, to be experienced by one person at a time on an intimate level, as can be the case in a private garden (figure 2-18).[9]

The concept of progressive realization, discussed earlier as a way of experiencing space, is particularly important in the design of an interesting vista. Because attention tends to be focused so directly on the terminus, the user's experience while moving through the space can be undeveloped and uninteresting if attention is not paid to what happens along the way. By using progressive realization, the designer can build anticipation and

change the focus from far to near, thus allowing different parts of the site to dominate at different points along the route (figure 2-19).

Whether the designer uses views, vistas, or a combination of the two in the visual design, substantiation of what the users see must occur in the plan function itself. For example, if the plan is designed to focus attention on a number of small group activities or on a detail at the building's entrance, but the design is handled so that a distant view of the horizon is dominant, the viewer's attention will be divided between what is going on close at hand (the small group activities or the details) and the more impressive distant view. A plan based on a dominant vista but in which a number of interfering functions take place will negate the vista's dominance and cause the focus to shift to the level of immediate experience.

2-18. A small-scale vista.

ORGANIZING ELEMENTS

Transferring the designer's mental image of the developed site onto the two-dimensional plan according to which the site will actually be built must be handled very carefully to avoid losing sight of the qualities that make the site unique. One plan element that is often used as an organizing feature, particularly in plans where the human factor dominates the natural, is the axis.

The Axis

The term axis is frequently interpreted to mean only something very formal, symmetrical, and straight. An axis is actually any linear plan element, whether curved or straight, that connects two plan elements or features. An axis dominates all other plan features, because the implications associated with connection are that the direction of movement and the entire focus of activities for the site will be along the axis. The movement is generally quite directional, oriented away from the use areas adjacent to an axis and toward a (usually) clearly defined endpoint of the axis. The tendency of the user is to leave the areas and follow the axis, either visually or physically (figure 2-20).

An axis may be symmetrical but more often is not (figure 2-21). It can turn, curve, bend, or be designed at an angle, but it cannot diverge along different paths. When divergence does occur, the single axis may break into a number of smaller, less dominant axes. A properly used axis is a unifying, ordering element and can be used to tie disparate sites together or to impose a certain discipline and rigidity on a single site. When axes are overused, however, the opportunity for users to choose among experiences is seriously limited.

A plan axis can be very monotonous if the design's symmetry and directional qualities are carried to an extreme. An unfortunately common example is a street system in which the axes' connecting features at either end are the same width, planted with equidistantly set trees, and lined with buildings of approximately the same character and usually the same setback from the street frontage.[10]

2-19. A vista using progressive realization.

2-20. An axis created by symmetrical lines of enclosing trees, terminating in a dominant vertical tower.

Hierarchy and Transition

The design of spaces employs a hierarchical system, composed of major spaces, minor spaces, and transition or connecting spaces. The design characteristics of the major spaces dominate the design; the designer's mental picture of these spaces is often the primary cause of trouble in transferring the plan to paper. The minor spaces take much of their character from the major spaces but may also serve as transition areas themselves, combining the design characteristics of two or more adjoining major spaces.

The transition or connecting spaces are the glue that holds the whole design together. The composition of these areas may be more difficult to resolve than the composition of major and minor spaces, since transition areas are truly compromise areas in terms of combining qualities and characteristics from a number of different site areas. Each transition zone may be designed in accordance with a certain set of constant qualities

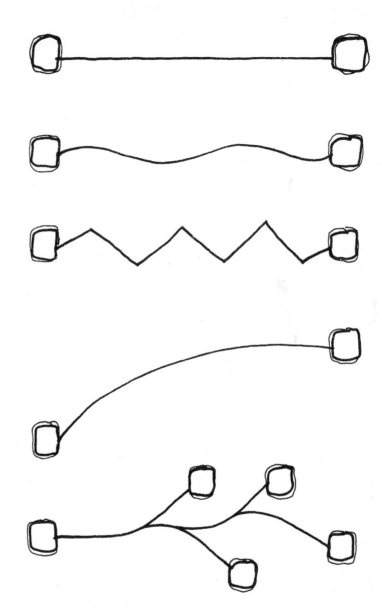

2-21. An axis can be straight, curved, angled, or bent, but it cannot diverge.

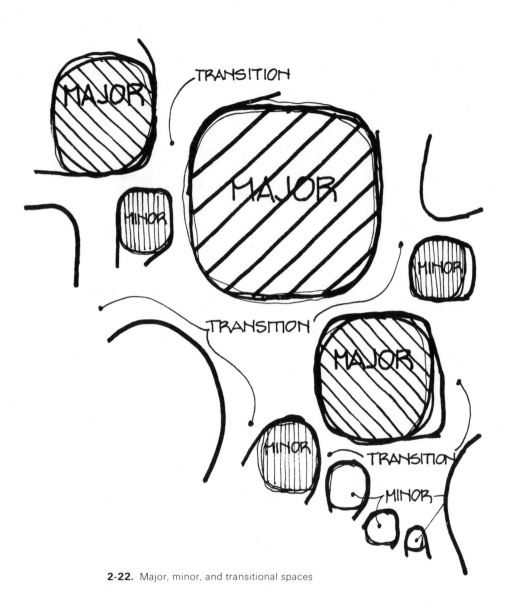

2-22. Major, minor, and transitional spaces

that are varied only slightly from zone to zone as the transition spaces blend with the major and minor spaces; or the transition may be more complete, sustaining freer, less thematic individual variation at the spatial interconnections (figure 2-22).

Scale Problems in Transition

Scale problems can occur in the transfer of the mental image to the physical design. This is most likely to happen when the dimensions of the plan as presented on paper are not synchronized with those of the actual three-dimensional project. A common example occurs in the design of planting plans based on a core of existing materials. Standard dimensions are frequently used to represent the canopy or spread size of various types of plant materials; and if the designer's mental image of the size of the existing plants matches the graphic plan symbols instead of the real sizes, the design may fail to establish enclosure or conversely may provide too dense a screen.

A similar effect can occur if the designer errs in estimating the mature size of the proposed plants. When this happens, the site may be dominated by towering trees or smothered by shrubs placed too close together.

A third problem of scale results from the disparity between the plan scale and the actual physical appearance of the site. The plan portrayal often creates the impression that more space is available than actually exists (this is because plans do not show the heights that give the site its feeling of enclosure). Visiting the site to get a total feel for the enclosure, designing by using three-dimensional sketches as well as plans, and even constructing models to show scale relationships can help avoid this problem.

BALANCE

Our first apprehensions of function and visual appearance in the world are based on an impression of balance. People and other living things rely on their own innate sense of balance to live, work, and move, and they associate the feeling of balance with comfortable spatial experience as well. There are two types of balance used in design: symmetrical or static balance, and asymmetrical or dynamic balance. Each has appropriate applications in the design of site, space, and structure.

Symmetrical or Static Balance

Symmetrical balance is based on the concept of symmetry, which is defined as objects or spaces in equilibrium around a given point. There are a number of different kinds of symmetrical balance, although natural symmetry over a plan-size area is very rare. The human form is an example of bilateral symmetry, in which objects or parts of objects are in equilibrium on either side of a center dividing line. The snowflake is a commonly cited example of radial symmetry, as are equilateral triangles, octagons, and an infinite number of geometric shapes (figure 2-23).

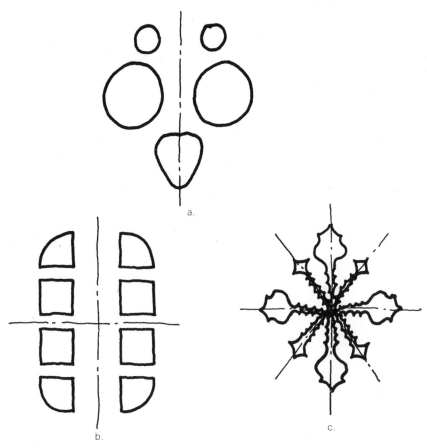

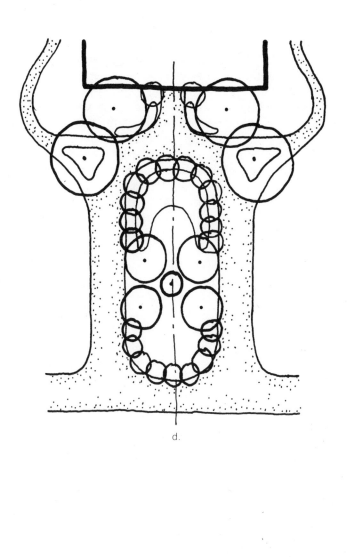

2-23. Symmetrical balance: a. bilateral; b. quadrilateral; c. radial; d. plan symmetry.

Symmetrical balance subjugates the meaning of each individual element to the design of the whole. A site and space in symmetrical balance must be designed so that the symmetry is apparent in its entirety from at least one point. If this does not occur, the user will be unaware that symmetry is present, and symmetry will lose its meaning as a functional organizer of the design. However, a danger of being able to see the symmetry in its entirety is that having once experienced it, the user may find that the balanced features of the space become merely repetitive. This accounts for the designation of symmetrical balance as static.

A symmetrical approach to site design is almost demanded by certain types of site/structure/space relationships. A building intended to present a strong, imposing, orderly image and itself designed symmetrically cannot be enhanced as well by dynamic balance as by designing the space and site to be in symmetry with it. Likewise, an axis intended to dominate the entire spatial experience by directing the viewer toward a terminating feature may require symmetrical balance to strengthen its power to direct attention as intended. The first purpose of symmetry always is to express the function that the design is intended to fulfill; symmetry cannot succeed simply by being imposed on any type of site and structure.

When elements are placed in a symmetrical relationship with one another, they set up a tension between themselves that forces them to be read as a unit. This makes the viewer aware of both sides of a building at once, just as the viewer perceives a whole person rather than just the left or right side of the body.

True symmetry is difficult to achieve in the design of exterior space because there are so many variables over which the designer has only minimal control. Foremost among these are the plant materials that form the backbone of many exterior spaces. Even though the utmost care may be exercised in choosing plants of matching shape, size, and growth characteristics, there is no guarantee that once they are installed those plants will continue to grow in exactly the same way and thereby maintain the original symmetry. For example, the soil may differ from one location to another, or the amount of light or moisture reaching the site may differ. One tree or shrub may bloom two or three days earlier than the others, or turn a different fall color, despite belonging to the same species.

Asymmetrical or Dynamic Balance

Asymmetrical or dynamic balance is based on the principle that the users of a space create their own perception of balance by associating forms, colors, lines, textures, and light in relationships of tension and equilibrium. The eye naturally seeks balance through the establishment of such relationships among whatever objects happen to be within the field of vision at a given time. This in itself has a dynamic or changing quality because each viewer establishes the balance in a slightly different way.

Dynamic balance is organic in form and is intended to change throughout the day or season. Its emphasis is on the plasticity of the site and space and on the effects that this changeability will have on the appearance of the structure. A space designed according to the theories of dynamic balance will be perceived as inviting and will be used by people only if there are no incongruities in the design. For example, a space whose forms are all balanced and appear to be set at equal distances creates an overall impression of imbalance and equal choice (figure 2-24).

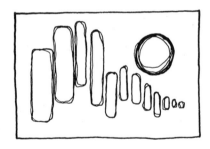

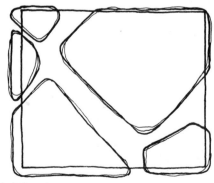

2-24. Dynamic balance.

The forms of the site elements and structural elements used in a project are key components in the creation of dynamic balance. Not only the forms themselves (whether rounded or square, organic or linear), but the actual size and mass of each form contribute to the impression of balance or imbalance. Objects or spaces of similar form but dissimilar size may be grouped so that one larger form balances with two or three smaller ones, using the space surrounding smaller forms to equalize the mass of the single larger form. Contrasting forms may also be balanced according to line, size, and relationship to surrounding spaces (figure 2-25).

Color has an important impact on the perception of balance. Since warm colors tend to stand out and cool colors to recede into the background, site areas that are balanced in form but imbalanced in color can indicate by their color selection areas of dominance and subordination—whether these relate to the overall balance being sought or not. Smaller areas of warm colors or less sharply defined shapes are needed to balance

larger areas of cool colors. Slight gradients of color may also change the apparent balance of a site. A common mistake in choosing site materials is to use too much contrast. Plants with variegated foliage are sometimes used as specimens without adequate neutral background to bear the visual contrast. A heavily textured, warm-tone aggregate paving, if used in combination with the cool colors of slate or the reds of sandstone, may appear out of place.

Both the quantity and the quality of light striking a site affect the balance perceived. Areas of dense shadow appear as large, heavy masses, in contrast with areas in full sun, which may appear two-dimensional and flat. The same expanding and receding effects discussed in relation to warm and cool colors are found when in relation to sunlight and shade: areas in sunlight appear to advance, while areas in shade appear to recede. Light and shade can be used in combination to create an illusion of greater depth in small sites or to focus attention on the foreground of large sites. As light fades in the evening, color intensity

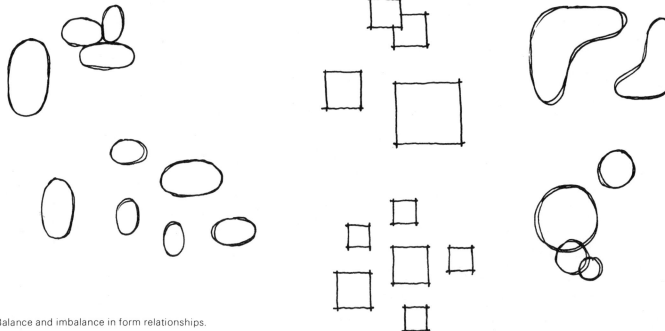

2-25. Balance and imbalance in form relationships.

fades also, causing objects to lose their color and shadow and their impression of depth.

The lines used in designing a site in asymmetrical balance are important not only for the forms they create, but in their own right as visual directors. A site composed entirely of strong horizontal lines with no verticals for contrast (or vice versa) may still appear balanced if the forms created by those lines fall into identifiable proportion to one another.[11]

Other Considerations in Creating Balance

The ground, overhead, and wall planes can be designed using lines varying from extreme boldness to timidity. The bolder and more dominant the ground plane line is, the more mass must appear in the overhead for balance. Strong vertical lines may need the addition of a longer, curved horizontal for contrast and balance. Short, sharp, jagged lines may require the use of a single bold straight line; undulating lines may be balanced by changing the depth and length of the undulations.

Texture comes from a word meaning "to weave," and texture essentially is a weaving of patterns into and over one another. Textures often occur in site design as a sort of overlay that provides either unity or contrast between adjacent forms. The careful use of texture can provide balance to otherwise unbalanced site areas, with small areas of coarse texture balancing large areas of fine texture.[12]

Consideration must be given to the seasonal appearance of site textures because many textures (those of plant materials, in particular) change over the year. A plant chosen for its fine texture while in leaf may be very coarse in its autumn fruiting or winter branching characteristics; or, contrarily, a coarse, large leaf may give way to medium-fine twigs. The change in texture is associated also with a change in density, and therefore in balance.

The distance at which a texture is observed and its distinctness individually or in a mass affect the apparent balance, too. The farther the viewer stands from a site feature, the finer that feature's texture appears. The more like-textured objects appear in a group, the more uniformity is created and the less dominant the texture appears by itself. It follows that disparate shapes and sizes of elements can be balanced visually through application of a uniform texture to them (figure 2-26).

By proper combination of form, line, color, texture, scale, and light, the designer sets the stage for the user to create a dynamic balance—one that may change on a daily or seasonal basis, but that certainly changes as different viewers associate the colors, forms, and lines in slightly different ways.

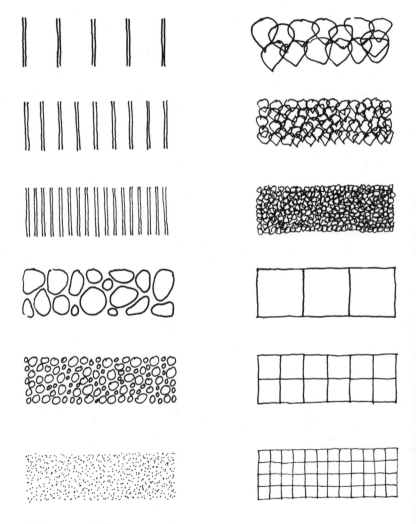

2-26. Textural differences.

CREATING UNITY

The designer must work with all of the aforementioned design characteristics in order to create unity and integrity of site, space, and structure. A number of other techniques can also be employed for the sake of unity, above and beyond paying attention to distance-height ratio, enclosure, viewing, and balance.

Transition Spaces As Unifying Elements

No matter how sensitive and interesting the design of individual spaces may be, if the experience of getting from one space to another lacks unity because of ill-designed transitions, the potential of the spaces will not be realized.

In the design of space and structure, the major transition spaces are those between the structure and the immediate built environment and those between the built environment and the unbuilt or natural landscape. Depending on the location and scale of the project, the latter transition may not occur. On rare occasions, a rather quick transition from structure to natural landscape may be necessary, but even in these cases the transition is first to more manufactured natural elements. An example of this is a building designed to be earth-sheltered or to fit into the native landscape, in whose transition space native plants are to be used. Even though the design moves in a short space from built to unbuilt, the internal and external materials must meet in that space and blend well together.

There was a period in landscape architecture's history (known as the "English Garden School") during which transition spaces between built and unbuilt spaces were almost totally ignored. This led to the design of massive, symmetrical, formal mansions floating in seas of well-kept lawns dotted with naturally placed shrubs and trees. At the other extreme are the often-advertised "foundation plants"—two of this at the door, three of that under each window, and a Christmas tree in the corner for good measure.

Structural Accents of Natural Features

To capture the feeling of bringing the outdoors in and sending the indoors out (or, in other words, to create smooth transitions between different types of spaces), the designer may choose to use a structural element to accent a preexisting natural feature.

For example, the highest vertical section of a building or tower may assume a position on top of the highest hill on the site, or a dominant natural element such as a boulder or rock outcropping may be reinforced by the designer's use of the same type of material in the construction. Another example of a structural accent is the use of a structural element as the backdrop for a natural feature such as a pond or lake. Conversely, of course, a natural feature such as mountains can be used as the backdrop for a structure.

Another way to form a smooth transition between built and unbuilt spaces is by treating a natural element architecturally. This may be done easily with plant materials, by clipping, or pruning the natural form of the plant to take on an architectural shape. Many buildings are surrounded by hedges that exemplify precisely this treatment. At the extremes of reshaping, plants may actually be pruned and trimmed to assume the shapes of spirals, ice cream cones, or even animals.

Natural, rough stones may be machined to have regular edges and smooth surfaces, enabling the material itself to continue to relate to the natural environment, after manufactured treatment. The milling of lumber is another example of architectural treatment of a natural element. Lumber may vary from very rough-sawn—indeed, almost hewn directly out of the trunk of the tree, to surfaced-four-sides and stained, polished, and sealed.

Interspersal of Natural and Man-Made Elements

A third method of creating smooth transitions is to intersperse the natural features with man-made or structural features. This is a common type of design in large park systems and other developments that need to combine areas of high-intensity use with more passive, natural areas. The constructed elements to be interspersed may be as simple as paved paths or freestanding benches and tables, or as complex as shelters or dwellings. Many successful high-density residential developments practice just such a technique, grouping units in eight- or twelve-unit buildings and then interspersing the buildings throughout the site, instead of clustering all the development in one portion of the site and leaving the remainder basically open and unused.

One caution about the use of interspersal to create transition is that smaller transition zones are formed at each one of the structural or man-made features being interspersed; the de-

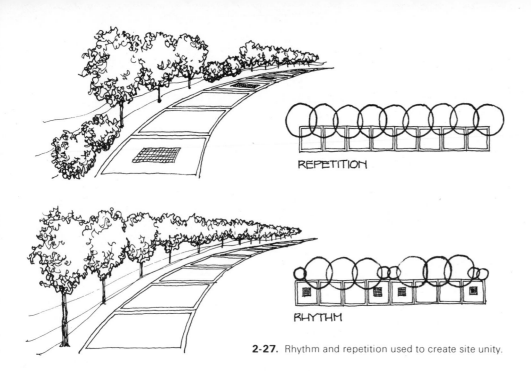

REPETITION

RHYTHM

2-27. Rhythm and repetition used to create site unity.

Contrast

The use of contrast to create site unity is a sophisticated and very successful design trick, if handled properly. Such handling depends on the designer's being fully aware of the design characteristics of the elements being placed in contrast with one another. Unfamiliarity with one or more individual characteristics can seriously weaken the planned contrast because of an unforeseen similarity between elements in that particular design characteristic, or because the contrast is so sharp that the elements read individually rather than as a unit.

A second important consideration in the use of contrast to create unity is domination by one of the contrasting elements. One element must clearly be dominant, and the other subordinate, in order for contrast to be present. If a designer working with color for contrast uses equal amounts of red and green, neither color will dominate, and no clear contrast will emerge because the user's attention will be divided equally. If, on the other hand, the designer uses eight or nine times more green than red, the contrast will be too great, and again unity will not have been created.

Contrasts in form, scale, texture, color, light and dark, line, and open and enclosed spaces can all be used in design. As long as diversity is developed alongside some form of order, the contrast will be successful.

It would be a mistake, however, to assume that several or all of the design characteristics of the elements of the site must be contrasted. Contrasting all elements can produce confusion about the proper hierarchy of the site's elements and features.

Rhythm and Repetition

Site unity can also be created by means of rhythm and repetition. Repetition is the recurrent use of a design element—whether it be the even spacing of plant materials or of patterns in concrete, or the echoing of colors and textures from one part of the site to another. A rhythm is set up when the repetition is interrupted by a different pattern, or color, or texture, or by a change in the spacing so that a new regular interval is established. If a single rhythm is carried on long enough, it becomes a repetition itself, and another variation will have to be introduced to reestablish a rhythm. The use of rhythm and repetition gives a cadence or beat to the movement through the site and produces a feeling of orderliness (figure 2-27).

signer cannot simply toss a handful of stones at the plan, put a bench or building at each stone, and expect the site to appear integrated. It is also important to vary the spacing, density, and balance of the features to provide a varied experience for the users. In other words, all of the design techniques that apply in the design of a single structure for a single site apply to interspersal as a transition technique as well.

Orientation

A fourth method of creating smooth transitions is to let the natural environment determine the orientation of the architectural features. This is easy to do when the view is dramatic—mountains or shore or a vast stretch of plains—but it takes more ingenuity when the views must be created along with the rest of the design.

In cases in which no exciting off-site features are present, the interest upon which orientation will be based must be created within the limits of the site. Japanese designers are masters of this technique—using a single specimen tree, an interesting landform just beyond the site's boundaries, or an open view of the rising or setting sun as a means of orientation.[13]

Similarity

Similarity can be used to create site-space-structure unity. A designer using similarity as a unifying element will not use identical materials, form, or lines in different parts of the design, but will use only slight variations—variations that are identifiably different but not so different that the points of similarity are not apparent. An example of the use of similarity as a unifying technique is in the use of vertical wood members for fencing, for screening the mechanical units and trash areas, and for barriers, together with the use of vertical lines in the mullions of the building, the lines of the walls, the trunks of the trees, and the light poles. The materials may vary slightly, as may the spacing of open to solid vertical members, but the proportion of open to solid and the overall impression of vertical line will remain essentially the same throughout the site.

Sequence

Sequence is a design technique that has been indirectly mentioned before in connection with how a person perceives spaces. Putting spaces in sequence means arranging them so that one space (and experience) leads logically into the next, without necessarily revealing the sequence in its entirety. The use of a sequence, depending on how it is detailed, can cause people to slow down and enjoy a small planting or site sculpture, or speed up to reach the next space in the sequence.

Common Scale

Common scale consists of the use of standard dimensions (for example, sidewalks 6 feet wide) throughout a site. It may also be based on a standard building material such as bricks or blocks or on a modular element such as a 12-by-12-foot deck area. A site design based on a module may include other elements whose dimensions are proportionate to the module, to add interest; in such designs, unity is preserved because of the common scale (figure 2-28).

Use of a Dominant Element

Site unity may be created by treating one site feature as a dominant element and designing all other activities and focal areas to be subordinate to it. This dominance may be exerted by a particular type of material (such as wood or brick), by a series of forms (such as horizontal terraces), or by an abundance of plants of certain types.

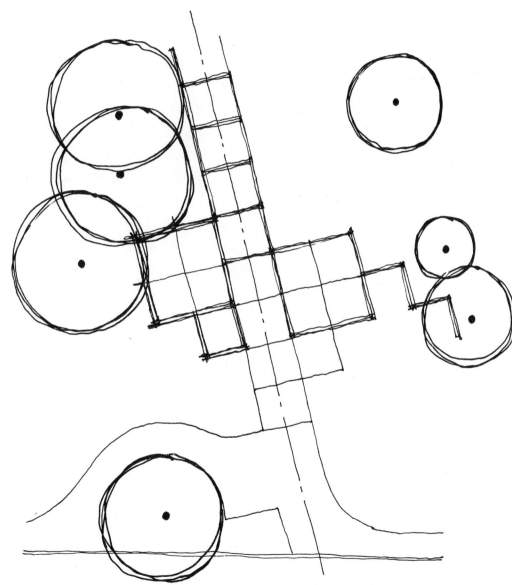

2-28. A square module (with varying sizes depending on location) used to give unity to a large site.

2-29. Water as a unifying element between the natural and built environments.

Since the function of the site and structure must be apparent and logical in the ground plane, the treatment of elements in this plane can be especially useful in creating unity. A particular paving pattern can be used throughout a site, varying as needed to respond to individual demands of scale and focus. Horizontality may be emphasized, by designing long, low-level changes in place of steep flights of stairs, or by orienting all site structures to the horizon line.

Water, which can be treated architecturally or naturally and can be directed vertically or horizontally, can be used as a unifying element in the ground plane with a considerable degree of success. Water appeals to people in all cultures and environments. It moves; it offers itself to taste and smell; it changes; it is liquid and refreshing; it can supply vivacity and excitement or restful background noise. When used in a flat sheet, water reflects its surroundings, and creates a reference plane in the ground surface. Because it can be used with equal success as a natural or structured element, water provides an ideal transition between site and structure. It can be used to separate use areas on a site physically while unifying them visually; and it can serve either as a detail element, in small subdued applications, or as a literal splash of excitement (figure 2-29).

Detail

Attention to detail deserves specific mention as the ultimate way of unifying structure and site with their surroundings. The treatment of details for purposes of aesthetics, future replacement (if necessary), maintenance requirements, and access determines to a great extent how and how much a site will be used.

Two major concerns in designing details for a site are the choice of materials and the installation of those materials. Either an inappropriately chosen material or a poorly installed material can reduce the visual quality of the site, increase the maintenance costs, and create safety hazards for the users.

A space designed with attention to detail will focus on how the details can best complement the various spaces and uses. All detailing should be chosen to encourage use of the areas designed for use and to discourage use of those not designed for it (without stating a big, flat "NO"). Unintended misuse or nonuse often results from badly chosen or badly installed materials.

A major detail, and one more likely to encourage use of a space than perhaps any other, is seating space. If the seating space offers the user a choice of sun or shade, privacy or an active location, it will encourage even more use. Seating allows people to talk, read, write letters, eat, or just relax in more comfort than if they have to stand or perch on a ledge or other structure not really designed for sitting.

William Whyte, in his "Nova" television special entitled "City Spaces, Human Places," estimated that one linear foot of seating per thirty square feet of area would produce adequate seating in public places. If more designers would analyze the need for seating on such a basis rather than concentrating on the strategic location of one or two benches (more as sculptures than as functional, sittable objects), more public spaces would be better used.

Seating is relatively easy to add to a space, but seating that is just added rather than being integrated into the overall design may be poorly located or may present maintenance problems such as requiring hand trimming. Benches that lack backs or are placed too close to one another will be uncomfortable. A backless bench requires its occupant to sit up straight rather than relaxing and promotes the uneasy sensation that one's back is exposed to whomever or whatever walks by. Shrubs or some other type of enclosure that screens the seated person on one or more sides will reduce the exposure problem, but not the actual physical discomfort. Benches that are too close to one another or that force total strangers to sit close will make people who wish to remain solitary uncomfortable.

In natural settings, designers often place benches that seem to have grown right out of the ground, giving them no introduction and nothing to rest on. While a hard surface under a bench is not absolutely essential, a walking surface leading to the bench would much more attractively invite passers-by to use it. Many an inviting, seating-height bench goes unused because it is removed from the main walking path.

The surfacing across which people walk to reach a space can deter use of the space, as can the lack of somewhere to sit once they arrive. While the use of an all-concrete site is barren, cold, and uninviting, surfaces used as substitutes for concrete should retain at least one of its qualities: they should be smooth and easy to walk on.

Many different applications of brick pavements are found in the landscape—from stepping stones laid in sand to carefully placed fish-scale patterns in mortar on a concrete bed. Brick on sand can be successful if the subbase has been compacted to reduce shifting due to freeze-thaw and moisture, and if the location is chosen carefully; however, people in a hurry, or in high-heeled shoes, will not be able to negotiate this kind of surface readily. Some types of brick (notably tile brick) are so highly glazed that their surface becomes dangerously slick when exposed to even the slightest amount of moisture. These belong indoors, not out.

Brick and other small modular paving materials, such as asphalt blocks of various colors or interlocking pavers, are widely liked as landscape materials for a number of reasons: they provide stable walking surfaces; they can be used in formal, geometric patterns or set at random; they can serve as transition materials between indoor and outdoor spaces, in the form of pavements, walls, or seats; and they are less reflective, softer to the eye, and more "of the earth" than concrete, all of which make them seem to belong in the ground plane.

Less stable pavements, such as crushed rock, gravel, or wood chips, can be used on a temporary basis or in locations that receive relatively little traffic. The use of these materials allows the designer to produce a gradual transition between high-intensity use areas and low-intensity use areas, without totally removing the walking surfaces.

The choice of surfacing should be made with full awareness of the degree of maintenance available and required. A plow blade can rip an uneven surface out, or chip a fine, smooth one beyond repair. A surface of grass pavers, intended to lose its distinct edges by blending with the surrounding grass over time, can be hard to mow and (if the pavers are set too far apart) almost impossible to walk on (figure 2-30). Figure 2-30 also shows the jarring note struck by a utility manhole, which is made to stand out even more by its apron of concrete.

All types of utilities can be eyesores as well as maintenance headaches if not treated as details in their own right. The use of trim or edging strips can often make maintenance easier when handled subtly, while lessening the visual disruption of the site. Screening trash sites and freestanding or obtrusive utilities with plant materials or fabricated fences and screens is a better alternative than just ignoring the utilities. Even signage, which often serves a very utilitarian purpose, can be designed with attention to color, location, scale, and choice of materials so as to blend with the rest of the design.

Light standards can be chosen to match the period of the design or to recede totally into the background except for the light itself. Since lights serve an important function in creating a

2-30. Hard-to-negotiate cobblestones and an obtrusive manhole.

feeling of night safety for the users of a site, the choice of fixtures and locations must be based on the light output. Considering the number of choices available, however, there is no reason to overpower a site with heavy, garish fixtures spaced too close together, or to pick lights that look more like little fireflies than anything functional.

Although a discussion about detailing for site design could easily fill a book by itself, at least one general rule can be stated: the choice of materials, scale, color, texture, and style, and the method of installing or locating the details should be made with the best possible use of the spaces involved, and the details should always be kept in mind—whether the design decision involves a single small planter to hold annual flowers, placed near the entrance to a private residence, or the choice of lights, benches, planters, trash receptacles, pavements, water, and signs for a downtown redevelopment.

REFERENCES

1. John Ormsbee Simonds, *Landscape Architecture* (New York: McGraw-Hill, 1961).
2. Yoshinobu Ashihara, *Exterior Design in Architecture*, rev. ed. (New York: Van Nostrand Reinhold, 1981).
3. Simonds, *Landscape Architecture*.
4. Simonds, *Landscape Architecture*, p. 155.
5. Simonds, *Landscape Architecture*, p. 122.
6. Simonds, *Landscape Architecture*, pp. 98–113.
7. Ashihara, *Exterior Design in Architecture*, p. 46.
8. Ashihara, *Exterior Design in Architecture*, pp. 43–45.
9. Simonds, *Landscape Architecture*, pp. 115–121.
10. Simonds, *Landscape Architecture*, p. 123.
11. Simonds, *Landscape Architecture*, pp. 131–138.
12. Harvey M. Rubenstein, *A Guide to Site and Environmental Planning* (New York: John Wiley & Sons, 1969).
13. Simonds, *Landscape Architecture*, pp. 70–75.

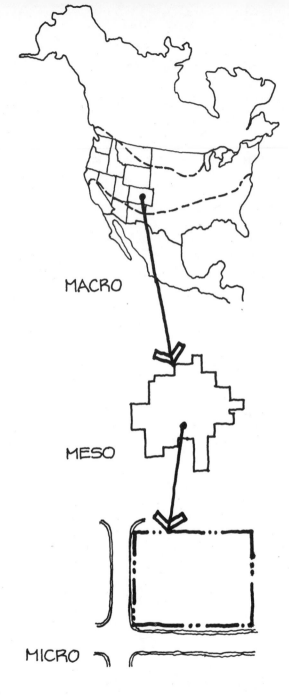

3 Climate and Energy Conservation

The same design characteristics and theories used in the design of exterior space and of transitions between site, space, and structure can also provide a basis for sensitive handling of climate-related environmental problems. Design to control climate and conserve energy while simultaneously offering an interesting environment and a variety of experiences for users can be accomplished if the designer understands the necessary elements of climate control.

LEVELS OF CLIMATE

The two main levels at which design for climate occurs are the macroclimate (the broad regional climate zones) and the microclimate (the small-scale, site-specific climate variations in those larger zones). A third level, the mesoclimate, applies to an area smaller than a region but larger than a single site. For example, Denver and Chicago are located in the same macroclimate, but the climatic characteristics of the two cities are quite different; each therefore has its own mesoclimate, and within each city specific sites have differing microclimates (figure 3-1).

Macroclimate

Macroclimate is determined geographically by a site's latitude, proximity to major bodies of water, and relationship to primary land features such as mountains. These factors affect the angle and/or intensity of the sun, the patterns of the prevailing winds, the level and form of precipitation, the duration of the seasons, and variations in temperature. The climate at the ma-

3-1. The three levels of climate.

croscale is controllable by the designer only on a very limited basis. There are four commonly recognized macroclimate zones: the cool zone, the temperate zone, the hot-arid zone, and the hot-humid zone. The climatic characteristics of these zones will be discussed later.

THE MAJOR CLIMATIC FACTORS

Climate at all levels in all zones is created by the interactions of four major factors, with significant influence from a fifth factor. The four primary factors are wind patterns, solar radiation, temperature, and precipitation. The interconnection of topography, the fifth influencing factor, with these four is indicated by the derivation of the word "climate" from a Greek word meaning "to slope."[1]

Taken separately, the impact of each climatic factor on the climate of an area seems rather simple to trace. However, when the interactions among factors are considered, the precise impact of each becomes far less clear. Understanding how climate affects site conditions and (ultimately) the design of the site, spaces, and structures requires introductory discussions of each of the climatic factors individually, followed by discussion of the interactions among them.

Wind Patterns

Wind consists of the movement of air. It is characterized by three variables: velocity or speed, direction, and degree of uniformity or turbulence. Wind is the climatic factor most influenced by topography, so a discussion of wind's characteristics necessarily covers its interactions with topography.

Velocity and direction combine to create prevailing wind patterns for a given site. The major global wind patterns arise primarily in response to the rotation of the earth and to topography. These affect the prevailing local winds by determining direction and season of flow. They also determine the presence of tornado- or hurricane-force winds during certain times of the year and of dramatic changes in wind direction (for example, those that bring warm, moist air into a dry region, such as the chinook winds of the Rocky Mountains).

Winds blow in three types of patterns: laminar, separated, and turbulent.

Laminar Winds

Laminar winds are layered, in a manner similar to a laminated beam. Each layer flows at a constant distance from the layers above and below it, and the speed and direction of the layers do not vary. Laminar winds are very predictable; because of this, they can be successfully dealt with in design for climate control (figure 3-2a).

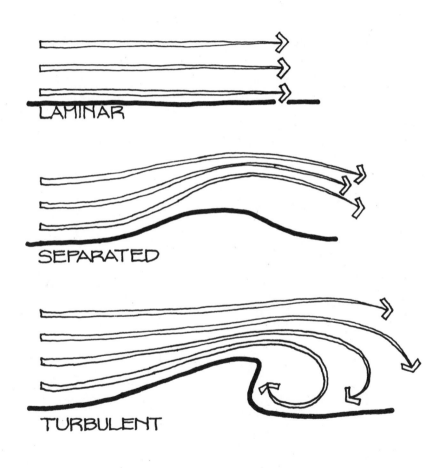

3-2. The three types of wind patterns.

Separated Winds

Separated winds are created when a difference in momentum occurs between layers of laminar wind. Such a change in momentum is due to a change in topography that causes the lowest layer to speed up (and separate) as it gets more tightly sandwiched between the ground and the air layer above it. As the bottom layer speeds up, it leaves a void behind it, and the pattern of the winds following changes to fill that void, creating turbulence. Separated winds are not as predictable or as controllable as laminar winds although to an extent the influences on separated winds of ground planes, plant materials, and structures can be determined (figure 3-2b).

Turbulent Winds

The third type of wind pattern, turbulent winds, is the most uncontrollable and the most unpredictable. Turbulent winds are separated to the extent that the wind is not layered at all, but instead blows in random directions at variable speeds. Turbulent winds are gusty, changeable, and influenced greatly by topographic and structural conditions. Many sites that had experienced only the typical wind patterns prevailing in their area have been subjected to frequent turbulence through additions of structures that created wind tunnels, through changes in grade that caused the winds to speed up or slow down, or through changes in surface materials that increased or reduced surface friction. Turbulent winds are characterized by dead spots and eddies, strong crosswinds and whirling gusts (figure 3-2c).[2]

Microclimate Wind Variations

Although the general pattern of prevailing winds for an area depends on the worldwide patterns, the changes that occur on a site-by-site basis are influenced by a number of other factors. Minor topographic variations cause many microclimate wind changes. Wind blowing across a flat site remains laminar and at full force. The profile of hills and valleys creates variations based on steepness and on the orientation of the slopes with respect to the prevailing patterns.

Because cold air is heavier than warm air, the airflow tends to be downhill during the night and uphill during the day. If the windward side of a hill is steeper than the leeward side, the change in wind pattern is more abrupt than if the leeward side is

steeper. Rolling hills break the wind slightly at each peak, and the wind speed in the valleys is reduced somewhat (figure 3-3).

The surface across which the wind blows affects the wind's force, path, and composition. A smooth surface offers nothing in the way of resistance; wind blowing across a smooth surface with an even topography will reach peak speed in a predictable direction. A very rough surface on the other hand, will break the wind at ground level, introducing separation or turbulence.

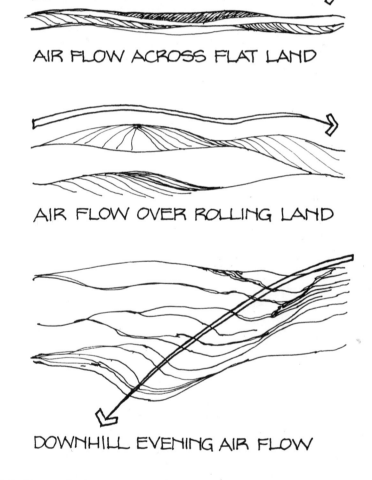

AIR FLOW ACROSS FLAT LAND

AIR FLOW OVER ROLLING LAND

DOWNHILL EVENING AIR FLOW

3-3. Changes in air movement are due to topography.

Wind brings with it moisture and precipitation, pollution or good smells from nearby locations, soil and debris, and trash from unguarded dump sites. Undoubtedly, many a person has vowed not to plant trees so as to avoid having to rake leaves, only to find that because of wind patterns, the treeless yard is full of leaves from a neighbor's tree, while the area under the guilty tree has blown clear. These site-specific variations in wind help determine the microclimate of the site.

The direction and speed of the winds in an area vary according to time of year. Fortunately, in most climates and at most sites, the north-to-south direction of the prevailing winter winds is nearly the 180° opposite of the prevailing south-to-north summer winds, making the task of designing to control wind year-round much simpler. Control of the winter winds is most important in terms of energy conservation (figure 3-4).

Solar Radiation

The amount of solar radiation reaching a site depends on the site's latitude and the earth's point of rotation on its axis; together these determine the angle of the sun in the sky. This angle is described in two different ways: as altitude (the angular distance of the sun above the horizon) and as azimuth (the angular distance of the sun from true north or true south), measured along the horizon in a clockwise direction in the northern hemisphere and in a counterclockwise direction in the southern hemisphere.[3] Sun angles change during the year, altering the amount of radiation reaching a particular site, and complicating design efforts to control and use the sun (figure 3-5).

The total annual solar radiation reaching a given location depends on the degree and the orientation of the site's slope. As it does in the case of wind, topography has a significant influence on the sun's effects. This radiation affects site conditions in four different ways: it may be reflected, absorbed, conducted, or moved through convection.

Reflectivity and Absorption

Different surface materials are able to absorb and to reflect sunlight at different rates. The reflectivity of a surface is measured on a scale from 0.0 to 1.0 called albedo. An albedo of 0.0 absorbs all heat and light, and radiates quickly. A material with a black matte surface is likely to have an albedo of 0.0. An albedo

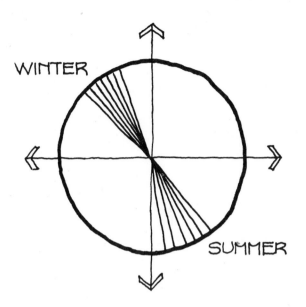

3-4. Prevailing winds are often opposite one another in winter and summer.

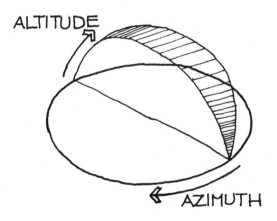

3-5. Sun angles.

of 1.0 is completely reflective, absorbing nothing. A mirror has an albedo of 1.0.[4]

The surface materials surrounding a site and on the site itself are thus able to influence how much of the sun's energy is reflected. The lighter and smoother the surface is, the more it will reflect. This may cause glare in the windows of a structure or hazardous driving conditions for drivers forced to face into the light reflected off a mirrored-glass building. Reducing the reflectivity of a surface by using darker, coarser materials such as grass and other groundcovers will also decrease the heat reflected into a structure or site area.

Conductivity

The reflective and conductive properties of a site combine to create a microclimate that is either stable or unstable. Conductivity describes the speed with which heat passes through a material. The drier and more porous or the lighter a material is, the lower its conductivity will be. Sites that absorb and release radiation slowly (sites that have low albedos and dense, solid surfaces) have stable microclimates as a result.

Convection

Convection also helps determine the relative comfort of a site. The most significant impact of convection involves the action of the wind in producing convective cooling. The convective exchange is based on a fluid-type movement of the air. The more turbulence is associated with the movement, the more heat will be dispersed. As air warms during the day, it becomes lighter and rises. This causes a flow of cooler air to fill the void left by the rising warm air; and as the air moves, a slight cooling breeze is created. The convective action combines with increased evaporation as the moving air sucks moisture out of the surroundings to make the temperature seem cooler.

Temperature

Temperature is described in two forms in relation to climate: as actual temperature and as effective temperature. Actual temperature is the dry bulb reading, uninfluenced by shade or sun, air movement, or precipitation. Effective temperature is the temperature the body actually perceives as a result of the combined effects of radiation, precipitation or humidity, and wind. This is the measure used to determine the comfort level of a site. The effective temperature may vary considerably from location to location, even on a small site, because of variations in the microclimate that affect the climatic factors. Such concepts as "wind chill factor" are used to assess the effective temperature.

A site's temperature is determined in part by the topography of the region, since temperature changes predictably as altitude changes. A decrease of one degree Fahrenheit for each 330 feet of rise during the summer and a change of one degree for each 400 feet of rise during the winter will occur as the air becomes thinner and less able to hold heat.[5] Such predictability is, however, contingent upon factors such as exposure to wind and daytime heat retention.

Precipitation and Humidity

Precipitation and humidity refer to the amount of moisture in the air at a given time and to whether that moisture is being held or released. The higher the vapor pressure (a function of the amount of water vapor in the air) becomes, the more uncomfortable people will be. As the water vapor builds and as the temperature changes because of wind and air movement, the air reaches a saturation point, and the vapor begins to fall to the ground in the form of rain, fog, snow, or drizzle (depending on the temperature).

Topography

The topography of a site affects the quantity of precipitation that falls or gathers and the relative humidity of a location; this is because topography affects the patterns of the winds that carry moisture. Small landforms receive relatively heavy precipitation on the leeward side of the hill (the side away from the direction of the prevailing winds), for the following reasons. As moving air begins to climb the windward side of the hill, the bottom layer of air speeds up and the top layers begin to cool. A sudden void develops beneath the moving layers of air as the topography drops away beyond the crest of the hill. This change in air pressure causes the air to drop its moisture. With more available space to occupy, the air moves less swiftly, and the moisture load carried by the air can no longer be sustained.

For land masses the size of mountain ranges and for landforms whose windward sides are much steeper than their leeward sides, the opposite effect occurs. As the warm, moisture-laden air begins to climb the steep windward face of the landform, it cools and finally reaches a point at which it cannot hold the water any longer. This accounts for the huge amounts of rain received seasonally on the windward side of the Pacific coast mountain ranges, as well as for the "rain shadow" on the other side of the mountain ranges. The light, dry air that finally surmounts the landform is compressed by the progressively greater air pressure from above as it flows down the lee side of the range, causing it to suck moisture out of the air rather than to drop it.

Wind flows from higher elevations through valleys to lower elevations in the evenings as temperatures drop. This produces a layer of cold air at ground level, which can cause fog and dew to collect at the lower elevations. The implications of such moisture collection with respect to the location of spaces for either early morning or late evening outdoor activities must be recognized by the designer (figure 3-6).

CLIMATE INTERACTIONS

The interactions among climatic factors produce what people perceive as the comfort level of local weather.

Temperature Reduction

The most important interaction between the wind and the other major climatic factors is its influence in reducing the effective temperature of a site through both convective and evaporative cooling. The more turbulent the wind is, the more cooling will result from convection.

In the process of evaporation, heat is removed along with moisture from objects and from the air. When no moisture is present, however, the wind makes dry air seem even drier. Consequently, the more surfaces available from which the wind can draw moisture, the greater the natural potential for cooling the site is. This shows up in the distinct temperature difference between urban and rural environments—the urban environment being warmer in both summer and winter because of the hard, moisture-poor materials that predominate in cities.

Man-made or manufactured materials are usually more reflective than natural materials, and they therefore cause more heat and light to be reflected into the site than do natural materials. In the rural countryside, the large amounts of vegetation and the rough, absorbent exposed soil allow a relatively large amount of moisture to be removed through evaporation and transpiration; these materials also conduct and store heat less readily than do city materials.[6]

Humidity Reduction

Wind can make the difference between comfort and discomfort when the air is heavy and humid, since it is able to remove humidity through evaporative and convective cooling. Although designers cannot control the effects of large, regional air masses (which are dictated by differences in air pressure), they can significantly influence effects in the zone closest to the ground, in which most human outdoor activities take place. Through channeling and other means of encouraging winds into a humid environment, this zone can be made more usable.

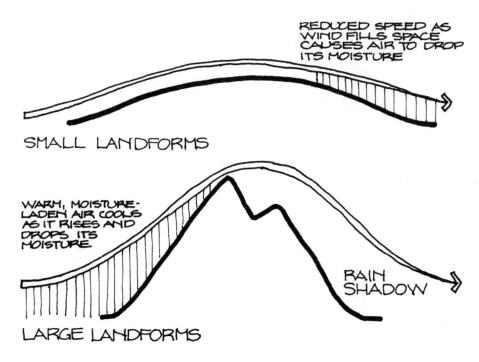

SMALL LANDFORMS

LARGE LANDFORMS

REDUCED SPEED AS WIND FILLS SPACE CAUSES AIR TO DROP ITS MOISTURE

WARM, MOISTURE-LADEN AIR COOLS AS IT RISES AND DROPS ITS MOISTURE

RAIN SHADOW

3-6. The effects of landforms on wind and precipitation.

Radiation Reduction

Unless the air is extremely dry, wind can also reduce the effects of intense solar radiation. Wind across a very exposed site can make the user feel as though the heat from the sun is not quite so intense. On the other hand, when the sun is hidden by clouds, a sharp wind will reduce the effective temperature of the site, and the users will not have the benefit of radiant energy from the sun that ordinarily might help counteract the effects of a damp, raw day and make the outdoor experience more bearable.

The amount of radiation present on a site has a direct effect on the temperature. Variations of as many as ten or twelve degrees can be obtained between sites located in full sun and those sheltered by shade. Wind and humidity, combined with the effects of radiation, do much to determine the comfort level of a site, but the most direct control over radiation occurs through the introduction of shade, either structurally or by using plant materials.

The sun has an effect on humidity, too. When the sun appears after a long rain on a warm day, the immediate effect of the evaporation of large amounts of moisture, combined with the sun's heat, is to transform the site into a steambath. In the winter months, direct radiation can cause ice patches and snow to melt rapidly, thereby increasing the usability of a site.

The Comfort Zone

The comfort zone is the range of temperature variation, humidity level, wind velocity, and amount of radiation within which human beings can work and live with minimal expenditure of energy and maximum comfort (figure 3-7). The breadth of the comfort zone varies with culture, age, and upbringing. A person accustomed to the tropics will be comfortable in a zone with a maximum temperature several degrees warmer than the maximum effective temperature comfortably tolerated by a person from New England. As people age, the lower limit of their comfort zone rises, and higher temperatures are required throughout the zone to maintain the same degree of comfort as before. There is also a gender difference: women generally prefer slightly warmer temperatures than men do. Upbringing and economic considerations also influence the comfort zone. The energy crisis of the early 1970s caused thermostats to be set lower in the winter and higher in the summer than they had been previously, and people over the last decade have adapted to a slightly broader comfort zone as a result. The limits of the comfort zone in the United States in the past were 69° and 80° Fahrenheit; following people's adjustment to the higher and lower thermostat settings, those limits are now probably closer to 66° and 83° Fahrenheit.[7]

THE GOALS OF DESIGNING FOR CLIMATE

In designing for climate, the designer's main goal is to create a comfort zone that allows maximum use of site, space, and structure.

Providing Shelter

Providing shelter is the primary means of creating a comfort zone on a site. The word "shelter" does not necessarily mean four walls, a floor, and a roof because that would imply that the only comfortable places to live and work are indoors. It does suggest that designers take all steps necessary to ensure that people be kept cool or warm, dry, protected from debilitating winds, and not suffocating in stagnant air pockets (figure 3-8).

Extending Livability

A second goal is to extend the livability of the site and spaces, so that the seasons of use are longer than they would otherwise be. Again, introducing shelter is the major means of accomplishing this goal.

Economizing

A third design goal is to economize on the means by which the site and (particularly) the structures are rendered functional. Sensitivity to climatic conditions and knowledge of the controls that can be designed for those conditions will enable the designer to reduce mechanical and monetary expenditures needed to keep the structures cool in summer and warm in winter, while making the site as safe as possible throughout the year.

SITE DESIGN FOR CLIMATE SENSITIVITY
Microclimate Analysis

The first step in designing site, spaces, and structures that are sensitive to climate is a thorough analysis of the site's microclimatic conditions. The analysis first addresses local climatological data, compiled on a mesoclimatic level. This will tell the

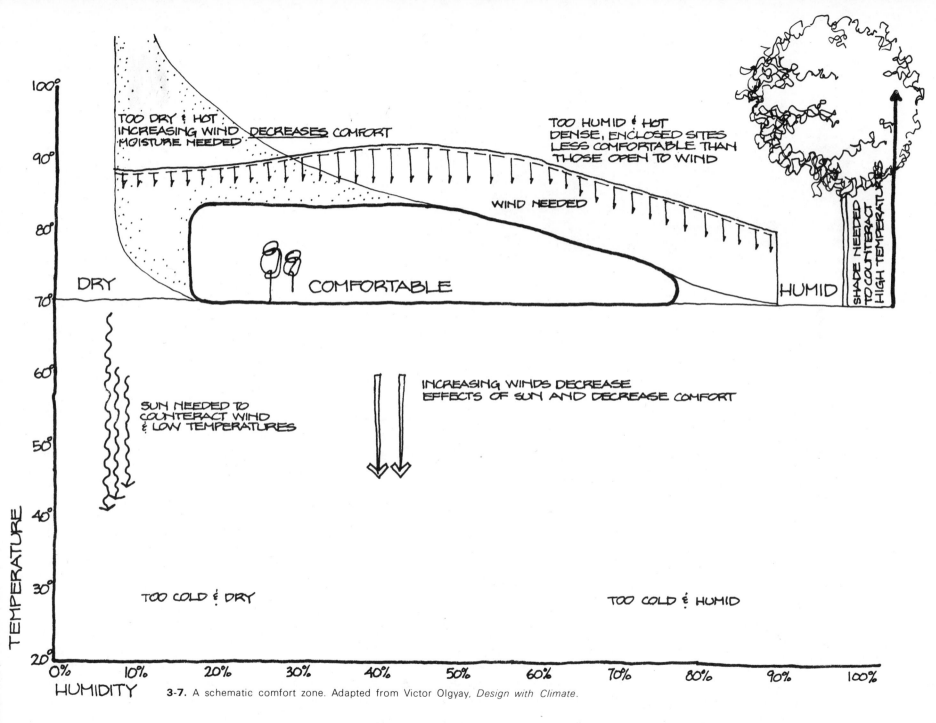

3-7. A schematic comfort zone. Adapted from Victor Olgyay, *Design with Climate*.

Text within the figure:

TOO DRY & HOT
INCREASING WIND DECREASES COMFORT
MOISTURE NEEDED

TOO HUMID & HOT
DENSE, ENCLOSED SITES
LESS COMFORTABLE THAN
THOSE OPEN TO WIND

WIND NEEDED

DRY

COMFORTABLE

HUMID

SHADE NEEDED TO COUNTERACT HIGH TEMPERATURES

SUN NEEDED TO
COUNTERACT WIND
& LOW TEMPERATURES

INCREASING WINDS DECREASE
EFFECTS OF SUN AND DECREASE COMFORT

TOO COLD & DRY

TOO COLD & HUMID

TEMPERATURE

100°
90°
80°
70°
60°
50°
40°
30°
20°

HUMIDITY
0% 10% 20% 30% 40% 50% 60% 70% 80% 90% 100%

3-8. Providing shelter from the wind, rain, and sun.

assessed, particularly for the region from twelve feet high down to ground level, where people will be most affected by the wind. Undesirable patterns created by off-site features must be managed as well as can be on site, through windbreaks and orientation of the structures and open spaces.

Using the known sun angles for summer and winter, the designer can analyze where the shadows will fall from each site feature. North-facing slopes and the north sides of buildings and other structures can provide desirable cool and shady outdoor activity areas during the summer months. Dense shade, such as occurs on the north sides, will create cooler, moister soils, limiting the types of plants that can grow there and leading perhaps to the design of comfortable oases in hot weather.

Topography, even at the microclimatic level, determines to a great extent the eventual distribution of the precipitation that falls on the site. Buildings and berms alike cast rain shadows. The cover of trees intercepts much of the precipitation that would otherwise fall to the ground. Small hollows and downhill slopes can become gathering places for fog and mist. Breaks in an enclosure can allow harsh winds to carry snow long distances into deep drifts. The overhangs, canopies, and closeness of adjacent structures can create places where little precipitation ever reaches the ground. An analysis of the site's topography, together with a knowledge of the amounts of precipitation and times of year when it falls, will allow the designer to anticipate the locations of damp, humid hollows (resplendent with mosquitoes and mosses) and of dry, summer-baked soils.

The temperature at the microclimatic level should be analyzed for variations between sunny and shady areas and between north- and south-facing exposures. The temperatures of south-facing, protected sites may be fairly constant—not fluctuating in response to winds that rarely penetrate, and warming and cooling more slowly. Small variations in position with respect to the wind, trees, or structures can significantly alter the temperature of a site.

Specific Site Variations

The microclimate is characterized not only by the interactions of the major climatic factors, but also by conditions existing on the site itself. These include the type and amount of groundcover—whether it is green and growing or paved—the quantity and placement of existing plant materials, soil condi-

designer the direction, duration, velocity, and variability of the winds, the angles of radiation, the number of sunny days that might be expected during the year, the timing, quantity, and form of precipitation, and the average temperatures. This information can then be applied to the site in question, on a microclimatic level (figure 3-9).

At this detailed level, the designer will consider the wind patterns across the site: how they are influenced by existing land or structural forms (either on the site or adjacent to it); whether wind tunnels or dead spots exist as a result of site conditions; what the wind carries with it (smells, moisture, dirt, dust); and the relative comfort the designer felt at various locations on the site when the wind was blowing.

In temperate and cool macroclimate zones, how and where the wind drops its snow load is important. The effectiveness of individual trees and groups of trees on blocking the wind and on controlling drift patterns of snow should also be determined.

In urban locations, the effects of the various building profiles and of the eddies and calm spots created by them should be

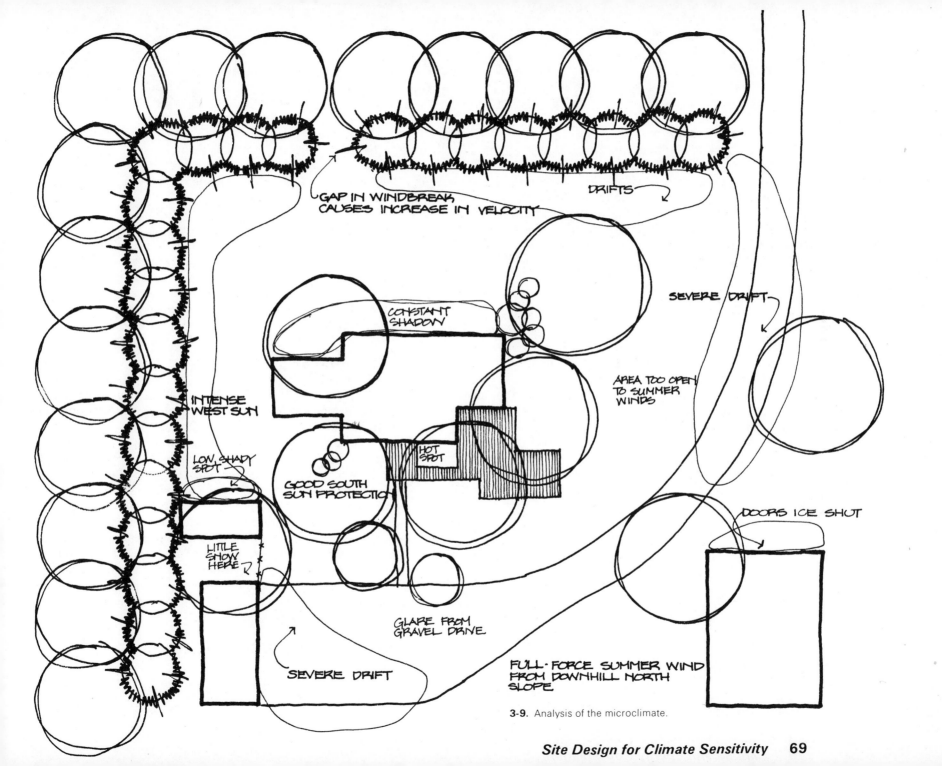

GAP IN WINDBREAK
CAUSES INCREASE IN VELOCITY

DRIFTS

SEVERE DRIFT

CONSTANT
SHADOW

AREA TOO OPEN
TO SUMMER
WINDS

INTENSE
WEST SUN

LOW SHADY
SPOT

HOT
SPOT

GOOD SOUTH
SUN PROTECTION

LITTLE
SNOW HERE

DOORS ICE SHUT

GLARE FROM
GRAVEL DRIVE

SEVERE DRIFT

FULL-FORCE SUMMER WIND
FROM DOWNHILL NORTH
SLOPE

3-9. Analysis of the microclimate.

tions, the orientation of the slopes, and the influence of off-site factors. All of these interact with the climatic factors to produce differences in shade, sun, wind, humidity, and temperature.

Type of Groundcover

The microclimate is influenced by the type of ground cover through reflectivity and conductivity (which affect the temperature at or near ground level), by the effects of the cover on relative humidity, by the ability of the site to shed or hold water, and by the types of matter that the wind carries across the site. The more natural the cover is (i.e., the greater the proportion of living plant materials that compose it), the more moderate the temperature fluctuations will be.

Plant Materials

Plant materials absorb radiation and offer surfaces from which winds can draw moisture through evaporation, thus cooling the site. A grassy area can be as much as 14 degrees cooler than bare soil; the temperature difference between grass and pavement can be even greater.[8]

Plants not only help to create the microclimate, they also are good indicators of the preexisting microclimate, as the designer begins analyzing the site prior to design. The vegetation that is native to an area is usually quite sensitive to soil conditions, air moisture, wind, sun, and temperature. If the plants growing on a site are known to be sensitive to air pollution, their health and vigor will attest to good air quality at the site. Plants that have been introduced, rather than growing naturally on the site, also give clues as to the site conditions, by how well they have adapted to their nonnative environment.

Vegetation helps to modify the microclimate by increasing the humidity; it offers the winds a multitude of surfaces from which water can be pulled by evaporation, and it provides moisture directly through the natural process of transpiration. Plants can intercept a considerable amount of the solar radiation available to a site, thereby reducing the heat at ground level. When laid out correctly, plants can reduce wind speeds by as much as 90%, compared to the same winds blowing in the open.[9]

Depending on the density of the cover, the darkness of the leaves, and the distance between plants, forested regions can absorb nearly all of the solar radiation falling on them. As sunlight penetrates through the forest canopy, it is filtered to a greater and greater extent, so that the sunlight falling on the forest floor is extremely weak. This principle of using trees to absorb solar radiation could be used much more than it currently is to reduce glare and heat in urban environments.

The ability of plant materials to block or channel the wind effectively is well known. The protected zone is a function of the height, penetrability, profile, and density of the materials used to create the windbreak.

Finally, the various surfaces of plant materials provide some control over precipitation by capturing much of the precipitation that would otherwise reach the ground when it strikes them. With the canopy holding moisture, less evaporates, and the soil meanwhile holds more of its moisture than it would if exposed to sunlight. The high humidity and low evaporation rate of areas situated under trees help to stabilize the overall temperature and microclimate of the site.

COMPROMISE

Producing the most sensitive design response to climate nearly always involves some compromise—whether to accommodate budgetary restrictions or to take advantage of special site factors such as an excellent view. At times, resolving the potential conflicts may be rather simple, such as deciding that providing a north-facing view of the city skyline is less important than achieving energy savings by means of a windbreak (particularly if the skyline view may eventually be blocked by development). The decision to plant a windbreak might be more difficult, however, if doing so meant imposing a separation of functionally associated areas of the property.

In order to make the necessary compromises, the designer must maintain clear communications with the client and with users of the site, and must exercise good judgment in weighing the pros and cons objectively.

THE FOUR MAJOR CLIMATE ZONES

Design for climatic conditions should first of all respond to the site's macroclimate. The climate of every site falls into one of four general types that exist in the world: cool, temperate, hot-arid, or hot-humid. Differences in the interactions among climatic factors produce the climatic differences that distinguish each of these four zones.

In all four climatic zones, the keys to successful design are control of the wind for evaporative cooling and control of the sun's radiation in order to maximize or minimize it. The optimum orientation for living spaces in all four zones is south, swinging slightly southeast or southwest.

Cool Zone

The cool zone is characterized by low or extremely low sun angles for a large part of the year, resulting in low temperatures and a short growing season. The frigid temperatures are often accompanied by large amounts of precipitation, usually in the form of snow. The air in this zone is often quite dry, notwithstanding the high annual levels of precipitation, and winds often drive the effective temperature during the winter down to −60° or −70° Fahrenheit.

In designing for the climate of the cool zone, it is necessary to maximize what little solar radiation is available, by orienting openings and collectors so that they face south or slightly east or west of south. In addition, heat loss must be minimized by such means as reducing the amount of surface or space facing into the winter winds, keeping openings (including windows) on those sides to a minimum, and using windbreaks, orientation, earth berms, and massive, heat-storing materials in construction to lessen the impact of the winds.

Temperate Zone

The temperate zone experiences overheating during some parts of the year and underheating during others; temperature fluctuations are large. The temperate zone's climate is more changeable than any other zone's, with distinct mesoclimatic variations within the zone. The variation in sun angles in this zone allows the designer to eliminate the sun from structures and spaces during the summer months and to admit it during the winter.

The direction of the prevailing winds is north-northwest during the winter and southerly during the summer, allowing wind control design that will prove advantageous throughout the entire year. Strong, gusty winds are prevalent in many locations, bringing blizzards during the winter and drying conditions during the summer. Relative humidity may be very high at times. The amount of precipitation, and how it falls, varies greatly across the temperate zone. This is the easiest zone for which to design, offering the greatest freedom to experiment with different site layouts and structural types while still responding to climatic conditions.

Hot-Arid Zone

The hot-arid zone is overheated during much of the year and lacks sufficient moisture for evaporative cooling to take place through wind and air movement. The design response to this macroclimatic zone attempts to minimize structural surfaces exposed to the sun's burning rays on both east and west faces by clustering and elongating structures in an east-west direction. Wind does little to reduce the effective temperature because available moisture is so low; consequently, the use of water features in this climate is a positive step toward introducing cooling—as is the use of shade. As in the cool zone, walls and structures should be massive, but their mass in this case serves to store coolness rather than heat.

Hot-Humid Zone

The hot-humid zone is also overheated, but in this zone there is too much moisture present rather than too little. Shade is necessary for climate control, and every small breeze must be summoned to assist in evaporative cooling. Rather than being clustered together for protection from the rays of the sun, structures may be separated to get as much air circulation to and through them as possible. In the hot-humid zone, outdoors moves in and indoors moves out fairly readily; housing therefore may consist of moving panels and walls, and patios and terraces may be roofed and screened without being sheltered from the wind.

WIND CONTROL
Orientation

The orientation of structures and sites in relation to prevailing winds can have an important heating and cooling impact. Since the winter and summer winds are generally opposite one another, a design that reduces the effect of winter winds (an important accomplishment in terms of energy conservation) can

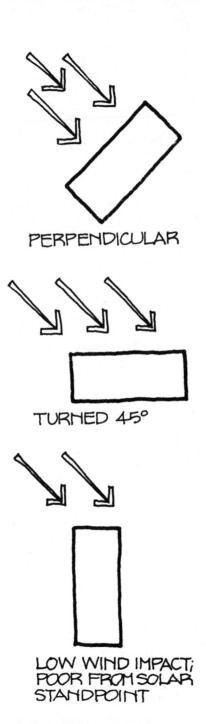

PERPENDICULAR

TURNED 45°

LOW WIND IMPACT;
POOR FROM SOLAR
STANDPOINT

also take advantage of the summer winds. A structure or space placed with its longest dimension perpendicular to the wind will receive the full brunt of the wind's force. Changing the angle to 45°—in other words, turning a corner into the wind rather than presenting to it the full surface of a side—can reduce its impact by as much as 50%.[10] However, if this angle is turned too far east or west, the efficiency of the structure's reception of solar radiation is reduced; therefore, a compromise is required between the optimum orientations for wind control and for solar control, and the design of other site features is also affected (figure 3-10).

Windbreaks

The primary control a designer has over the wind is by use of a windbreak composed of plant materials, structural materials, or both. For greatest effectiveness, a windbreak should be placed perpendicular to the direction of the winds needing control, and should extend beyond the zone needing protection in both directions. This is because the speed of the wind when it reaches the end of a barrier and begins to flow around it unimpeded once more is actually greater for a short distance than it was before it reached the barrier. A solid barrier will form a protected zone that is shorter than that created by a penetrable barrier, and the velocity reduction will be greater but for a shorter distance (figure 3-11).

The ability of a windbreak to provide protection is directly related to its location, density, and composition.

Windbreak Locations

Location must be considered from two standpoints. First, the most effective windbreaks are those placed perpendicular to the prevailing winds. Second, the distance of the protected zone is directly related to the height of the windbreak. The shorter the windbreak, the shorter the zone of protection.

A windbreak actually reduces the speed of the wind not only behind it, but in front of it as well because the wind begins to slow down and pile up on the windward side of a break just before it changes direction and flows over the break. The length of this area of reduced wind speed on the windward side of the barrier may be from two to five times the windbreak's height; the length

3-10. Influence of turning a building various ways into the wind.

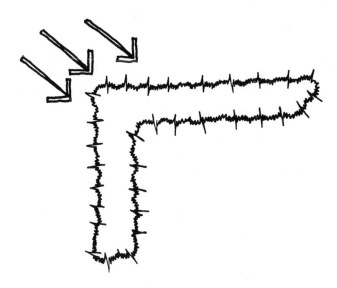

3-11. Windbreaks should be perpendicular to the prevailing winds.

of the control zone on the leeward side may be as much as fifteen times the height of the barrier. Any site features or structures needing protection should be located within that protected zone (figure 3-12).

Degree of Control

The actual reduction in velocity is determined by the profile of the windbreak and by its density. The more penetrable the windbreak is, the longer distance the protection zone will extend on the leeward side, and the lower the actual reduction in velocity will be. Optimum penetrability is about 50%, which is not hard to accomplish if plant materials are used. The more penetrable the windbreak is, the less difference in pressure from one side of the windbreak to the other will result and the less dramatic the shift in direction and velocity will be.

On the other hand, a solid barrier reduces the wind velocity on the leeward side to such an extent that it drops to nearly zero—but the original speed is resumed much more quickly than would happen if a penetrable barrier were used, and considerable turbulence in the zone immediately adjacent to the dead spot may also occur. A solid barrier creates a relatively great difference in air pressure between the windward and leeward sides, which in turn reduces the size of the protected zone.

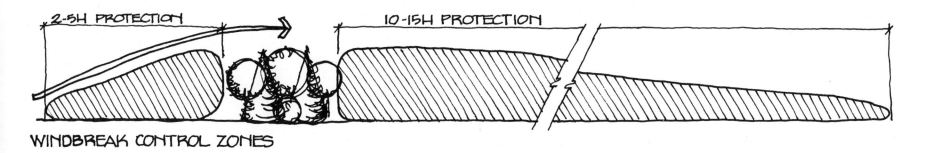

WINDBREAK CONTROL ZONES

3-12. The wind control zone established by a windbreak.

PROTECTED ZONE

DEAD AIR ZONE

TURBULENCE

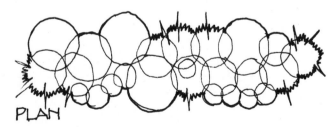

PLAN

LESS PENETRABLE, MORE UNIFORM WINDBREAK

Windbreak Design

The optimum design for a windbreak is perpendicular to the wind with a composition that permits 50% wind penetration, and with the sites and structures placed so as to lie safely within the protected zone on the leeward side (figure 3-13). The structures should not be situated too close to the windbreak because there is a dead air pocket where little air movement occurs, just to the lee of the break; neither should they be situated too far distant, where the velocity reduction is no longer significant. Although the protected zone may extend to a length of between ten and fifteen times the height of the windbreak, optimum protection occurs at distances from the barrier of between five and seven

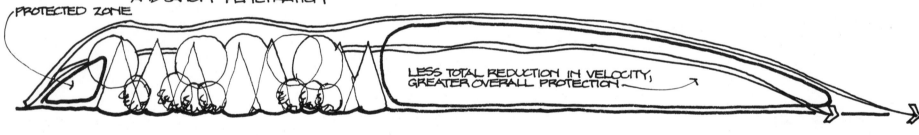

WIND SPEED REDUCED BY ROUGH PROFILE AND CANOPY PENETRATION

PROTECTED ZONE

LESS TOTAL REDUCTION IN VELOCITY, GREATER OVERALL PROTECTION

PLAN

MORE PENETRABLE, LESS UNIFORM WINDBREAK

3-13. Comparison of more and less penetrable windbreaks, and their relative degrees of protection.

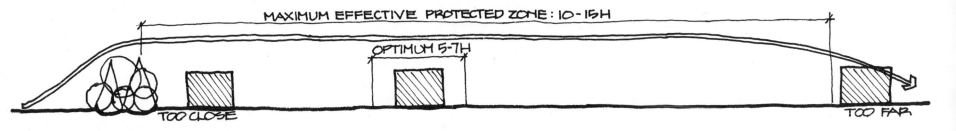

3-14. The area of optimum protection behind a windbreak.

times the height (figure 3-14). The areas to be protected must not be situated too close to the ends of the windbreak, either, because the wind, as it rounds the ends, will actually increase to a slightly greater speed than it had when moving freely across the land—before it reached the windbreak (figure 3-15).[11]

Not only the density, but the heights, the uniformity of canopy, and the materials used affect the efficiency of a windbreak. An obvious difficulty lies in making plans that rely on a windbreak for immediate protection, when plant materials may take years to reach their mature heights. An acceptable compromise may be to use a few larger materials intermixed with smaller ones or to locate the areas to be protected slightly closer to the windbreak than is actually desired, and then to remove one or two of the innermost plant rows as their height increases.

A windbreak of completely uniform height is not highly efficient in reducing wind velocities; additional speed reduction can be accomplished through increased friction and small air pockets if the heights along the top of the windbreak vary slightly. When the heights vary, the wind assumes a roller-coaster type of movement, slowing down as it hits each change in the profile.

The density of the materials used affects the windbreak's penetrability, as does the approximate region of the densest portions—in the upper canopy, at intermediate heights, or close to the ground. If the windbreak is uniformly dense at all levels, the zone of protection will be efficient, requiring no specialized treatment to control unusual wind patterns such as are created by windbreaks that are not uniform.

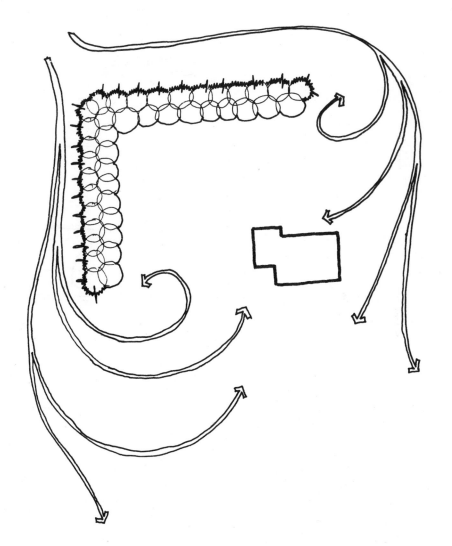

3-15. Placing a structure too close to the ends of a windbreak reduces the protection.

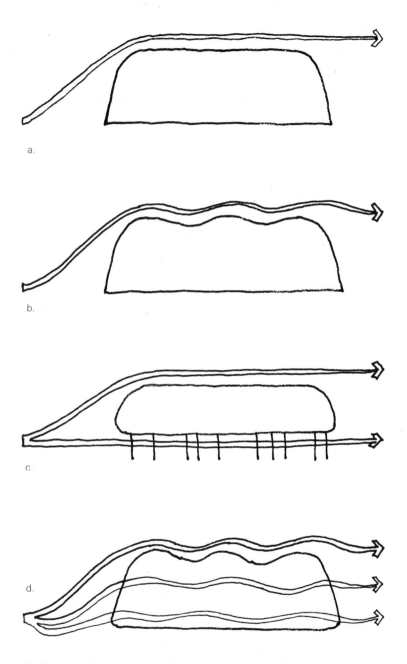

3-16. How different profiles affect the wind: a. too uniform; b. varied profile, too dense; c. canopy; d. varied heights, penetrable.

Since most human activities take place at or slightly above ground level, the ground-level area needs the most protection. As windbreaks composed primarily of overstory trees and evergreens mature, they tend to lose their lower limbs, creating a canopy of foliage supported by bare trunks. Ground-level wind velocities actually increase in such cases, as the wind is squeezed between the lower edge of the canopy and the ground. This can be avoided, however, if the designer uses understory and shrub materials in the windbreak to present an evenly distributed barrier.

The use of too many evergreens can decrease the windbreak's penetrability, thereby decreasing the length of the zone of protection. Thus, an optimum windbreak will combine plant materials of various heights and densities, including both deciduous and evergreen species (figure 3-16).

In cross section, the windbreak should be rectangular, with an irregular surface. This is more desirable than any type of pitched-roof cross section, which works to create either an updraft that allows the wind to return quickly to its previous speed or a downdraft that directs the wind toward the zone to be protected rather than away from it (figure 3-17).

Estimated fuel savings possible through careful planning of climate control for a site by means of windbreaks range as high as 30%. Considering the minimal investment in materials and time needed to establish a windbreak, this represents a great savings.

While reducing the force of winter winds is important in the cool and temperate zones during at least half the year, channeling the beneficial summer winds is also important in these zones and is extremely important in the hot-humid zone at all times.

Airflow is from zones of high pressure to zones of low pressure; a solid structure or other barrier creates a high-pressure zone on its windward side and a low-pressure zone on its leeward side. Providing openings in the structure or between plant materials to allow airflow from the high-pressure zone to the low-pressure zone can therefore create positive circulation through the spaces. Large openings should be on the leeward side, so that the winds will speed up as they go through the zone that is to be cooled. The use of a single tree or hedge can effectively channel winds into spaces that require them, without interfering with the protection of the spaces during other times of the year. The same principle can work to the disadvantage of the design if it occurs within a windbreak: the wind will seek any outlet available and will speed up as it passes through an opening.

Although wind protection can result from sensitive placement of even a single tree, the use of at least three rows of vegetation is recommended. This allows the plant materials to be staggered so that the wind is blocked at all points. If space for three rows is not available (as in urban areas), whatever protection can be provided will help. Plants can also form an insulating space when placed close to the walls of a structure, allowing still air to act as a barrier against the invasion of the wind.

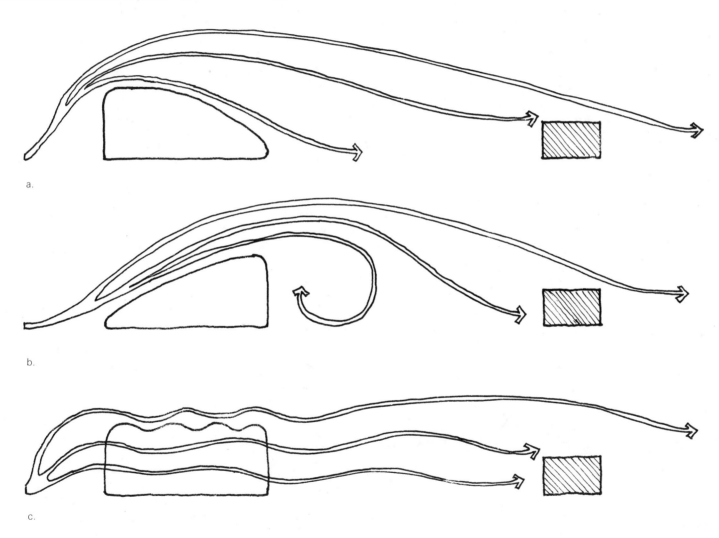

3-17. Cross-sectional differences in windbreaks; a & b. pitched roofs; c. irregular but rectangular.

Topographic Control of Wind

Spaces and structures can be situated so as to take advantage of the influence of topography on wind patterns. This is a design area in which much compromise is often necessary.

Because of the common pattern of airflow from high elevations during the day to lower elevations at night, circumspection must be used in locating anything at the lower end of the airflow if avoiding relatively moist and cool evening conditions is desirable.

A structure located at the highest available elevation receives the full force of the winds in all seasons; depending on the difference in elevation between the high point and the windward side of the hill, it may take years for a natural windbreak to reach sufficient height to block much of the wind. This may, however, be the optimum location for sites in the hot-humid zone because it enables them to take advantage of as much air movement as possible. In general, a site partway up the leeward side of a hill has a more stable microclimate than one at the top or bottom of the hill, and is more amenable to activities at all times of the year.[12]

Compromise may be necessary to provide the most efficient access roads, or to take advantage of a particularly interesting view. Depending on the site's orientation, the placement of a windbreak could dam the flow of air from high elevations to low elevations, causing cool, damp conditions to persist in what should have been a protected zone (figure 3-18).

SUN PROTECTION

Design control of the amount of solar radiation reaching a site must be based on three factors: orientation, slope direction and degree, and composition of the surface and other materials that will allow either reflection or absorption.

Orientation

The importance of orientation of site features has already been discussed to an extent. In all climatic zones, a south orientation for the major outdoor use areas and openings is desirable. The optimum orientations for each of the four zones are given in figure 3-19.

Degree of Slope

The degree and direction of slope affects the amount of direct solar radiation received on a site because the largest amount occurs at an angle of 90° to the surface being struck. South-facing slopes, therefore, receive larger amounts of solar radiation than any other orientation, except when the sun angle is low. At these times, the radiation received can be greater for a west orientation.

The relationship between orientation and degree of slope has implications for the growth of plant materials. Soils and air warm more quickly on the south, producing as much as a two-week lag

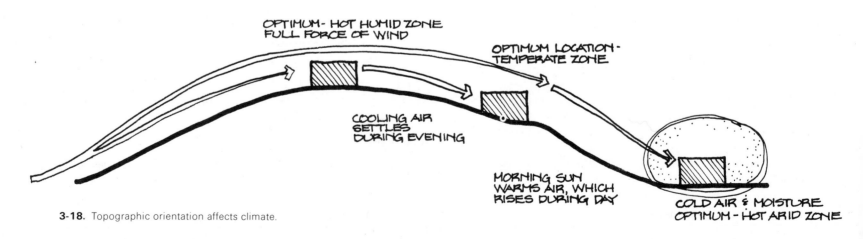

3-18. Topographic orientation affects climate.

in plant development between north and south orientations during the spring and the fall for the same site. If radiation reduction is desirable (which it is in all zones except the cool zone for at least part of the year), the designer should orient spaces and structures to avoid a heat buildup through the day, as it is this buildup—and not the high temperatures that are reached for only a portion of the day—that causes uncomfortable conditions.

Shade and Reflection Reduction

The major control of solar radiation occurs through introducing shade and at the same time reducing reflective surface area. Tree cover, grass, and other groundcovers play an important part in reducing the amount of radiation reaching a site, intercepting as much as 90% of the direct radiation—which in turn keeps the ground temperature from rising and eliminates much of the heat and light reflected into the structures or adjacent spaces. How effective plant materials are in reducing the need for artificial cooling depends on the types of plants chosen and their placement.

Plants for Shade

The primary bases for choice are a plant's ultimate height and spread, its shape, and its density during different seasons (assuming that the plant has already been chosen with regard for hardiness). Often a canopy tree, chosen and placed specifically to shade a building or space during the warm afternoon, fails to do its job because the designer forgot that as a tree matures, its limbs grow up almost vertically. When a mature tree's shadow lengthens in the afternoon, the only shade it will throw on a building it is too close to will be from its trunk. Using a combination of large and small plants is a sensible way to avoid this situation.

The shape of the plant determines the shadow pattern it throws. A building or space requiring almost total shade will benefit from planting wide-spreading, rounded trees or trees with a vase shape such as that of the old American elms. Shade may be desirable only on a narrow part of the site, in which case a more narrow, upright-growing plant might be used.

The density of the plant determines the deepness of the shadow it throws. A very fine-textured tree will throw a shadow

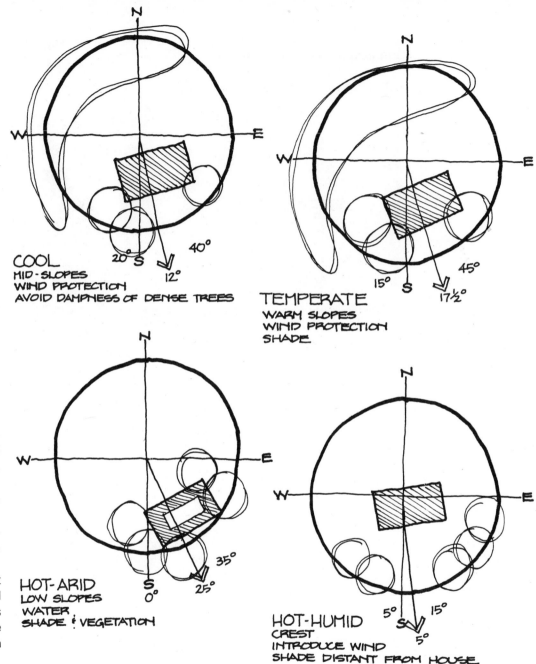

3-19. Optimum orientations for solar radiation control in all zones. Adapted from Victor Olgyay, *Design with Climate.*

3-20. Overhead structures are transitional between spaces and provide shade.

that barely reduces the amount of light reaching the ground or building surface. On the other hand, some plant materials are so dense that almost no light penetrates them; these very effectively limit the radiant energy reaching the space or structure, but often allow nothing to be grown under their canopies.

Density may also change with the season of the year, which is an advantage in zones where shade is desirable for part of the year but direct solar radiation is needed for heating at other times. Few deciduous trees remain as dense in winter twig as they are in summer leaf, although some thicket-formers approach a reasonable shading density. Dense evergreens, how-

ever, may be too dense to use in cool-zone climates. The types of plants used for control of solar energy should be dictated by the specific site conditions.

Constructed Shade

Shading can also be accomplished through the use of mechanical or constructed devices such as canopies, lattices, awnings, and reflective or treated glass. Partially open overhead structures serve the dual purpose of supplying a gradual transition from totally enclosed structure to totally open landscape and providing needed shade (figure 3-20).

Choice of Materials

The choice of materials plays a significant part in reducing or increasing the amount of heat and light reaching any part of the site. It may be desirable to use highly reflective materials to introduce as much radiation as possible to the structures or spaces during the winter months and then to reduce the effects of those reflective materials through shade during the summer. An overabundance of materials with a high albedo will tend to create a microclimate that is subject to relatively large fluctuations in daily temperature.

PRECIPITATION CONTROL

Control of precipitation at the microclimate scale depends in part on proper orientation and location of the openings into structures, of the spaces that are intended to be used outdoors, and of vertical and overhead protection. The relationship of precipitation to topography has been mentioned. Although topographic changes at the microclimatic level are usually not severe enough to cause obvious rain shadows, there can be a slight difference in the total amount of rain or snow falling on the leeward and windward sides of hills. Valleys and low depressions are subject to the collection of dew, fog, and frost, making them more difficult to use on a twenty-four-hour basis and limiting the plant materials that can be successfully grown in them. Regrading to minimize such low pockets can help reduce the number and size of locations where precipitation in this form will gather, extending the usefulness of the site.

Orientation and Location

If the hardest rains of the season are borne by southwest winds, the designer would be wise to locate screened porches and patios in a protected, east-northeast location, if this is compatible with other site conditions. This arrangement provides an area with overhead protection, open sides, and an attractive orientation: some of the most pleasant site experiences occur while watching the rain and feeling the coolness it brings without getting soaked to the skin. Orientation of inside-outside areas also has implications for extending the variety of furnishings and materials that can be used without being damaged. Houseplants or other exotics may be successfully located in such areas, too, out of the path of the prevailing storm winds.

Plant Materials

Plant materials play an important part in the control of precipitation, intercepting and holding much of the moisture that would otherwise reach the ground and later releasing it slowly, either to the air (in the form of evaporation) or to the ground. The denser and more solid the foliage mass is, the less moisture will reach the ground—shielding anyone standing beneath the trees from getting wet, but withholding water from anything under the canopy that may need it to grow. Evergreens capture and hold more moisture (particularly in the form of frost or fog) than deciduous trees because of their pointed needles and because of the sharp angles at which the needles join the twigs. However, this moisture, once held, can also be released quickly in a large mass, as when a stand of evergreens suddenly dumps its entire load of snow on an unsuspecting head.[13]

Snow Control

Control of precipitation is most necessary in the cool and temperate zones, where plants and structures alike are used to control snow drifting. Although topography determines to a great extent where snow will fall in the first place, modifications in topography (because they affect wind patterns) can reduce or even completely eliminate severe drifting in an area. Unfortunately, some of the worst drifts occur because of a difference in air pressure between one side of a building and the other, causing the snow to be dumped on the driveway or in front of the doors; little or nothing can be done about this because of the location of the building and the openings into it.

Certain topographic conditions result in predictable drift patterns. Notable among these are sharp dropoffs, such as those that occur along the sides of many roads. Since the wind slows down when it encounters an increased volume of space through which to blow, it releases its load as it reaches the lip of a ditch. The length of the tail end of a drift depends on conditions of adjacent topography and on the speed of the wind. Hilltops are frequently scoured completely clear of snow, with the load being dumped on the leeward side as the elevation drops away. Site features that must remain accessible at all times, as well as access roads to them, must be located with these considerations in mind.

Snow fences are oriented as windbreaks are: perpendicular to the direction of the prevailing winds. Slight modifications in this orientation are possible if sufficient space exists between the feature being protected and the fence. The height of the drift is a function of the height of the snow fence; shrubs and low fencing are more effective in controlling the drift pattern than are tall trees. Most constructed snow fences are four feet high, but natural or living snow fences may be slightly taller. As with windbreaks, the most control (in this case, control over the largest volume of snow) will occur if the snow fence is about 50% penetrable. A solid barrier produces drifts on both the leeward and windward sides that are short, deep, and nearly equal in size, but it holds relatively little snow in comparision to a permeable fence. A permeable fence produces a short, shallow drift on the windward side, as the wind breaks its speed slightly upon encountering the obstruction, and a long, shallow drift containing a great deal of snow on the leeward side.

Because of the many variables that affect the overall length and deepness of the drift pattern, the formulas used for calculating the best location for the barrier yield approximations at best and contain "fudge factors" of up to a third of the total drift distance. A general guideline to use is that a barrier 4 feet tall and 50% penetrable should be located at least 60 feet away from the area to be protected from drifting; 70 feet is better. For a 6-foot-high barrier, the location should be at least 70 feet back.[14]

There are advantages and disadvantages to both constructed and living snow fences. Constructed snow fences must be erected and torn down each year, the fencing materials often need repair or replacement on an annual basis, and special equipment is needed to drive the posts. From an aesthetic standpoint, the standard lath-and-wire snow fence leaves a great deal to be desired.

Living snow fences must go through an establishment period before they are really useful as snow barriers, and this period involves planting costs, initial maintenance costs (watering and fertilizing are rarely done in most areas thereafter), and occasionally replacement costs. Once the barrier reaches its optimum height, it must be held there by manual pruning unless the plants were selected—as they should have been—for their proper mature height. Roguing out unwanted trees and shrubs can become a problem in high-maintenance areas. Once the living snow fence is firmly established, however, nothing more need be done to it, and the long-term investment thus drops to virtually nothing. Certainly, from an aesthetic standpoint, the living snow fence has more to offer than does the standard constructed fence.

TEMPERATURE CONTROL

Control of temperature at the microclimatic level requires attention to all climatic factors. The influence of wind in providing cooling through evaporation and convection has been discussed, as has the effect of the sun in providing radiant heat. The cooling effects of plant materials, too, can be important in creating a comfortable microclimate. Not only does the canopy lower the temperature beneath it (by reducing the amount of solar radiation there), but the absorption of sunlight (and solar radiation) by the dark green leaves causes additional decreases in reflected heat. The combination of the darker, less reflective surfaces of plant materials with the ability of such materials to absorb radiation and cause cooling through evaporation can cause a temperature difference of as much as fourteen degrees between bare soil and a grassed surface.[15]

WATER AS A CLIMATE MODIFIER

The presence of a body of water on or near a site helps moderate the site's microclimate. The mass of water acts as a heat reservoir, warming up gradually during the spring and remaining at a reasonably constant temperature throughout the warm season. Except when the sun is low in the sky, the albedo of water is very low, causing little reflection to surrounding surfaces. The winds moving across the surface of a major body of water blow inland during the day and in the opposite direction at night. When the air temperature is very high, even the slightest breeze across water will produce evaporative cooling and make the weather more bearable.

The position of the site with respect to the water greatly affects the water's influence on the site's microclimate. Sites located on the windward side of a body of water will not receive the beneficial cooling effects of the wind blowing across the water. Sites located on the eastern shore will be subject to intense glare from the reflected western sun as it sinks. Sites located on the western shore will be somewhat protected from the impact of the prevailing wind blowing across the open water or ice in the winter and will receive the cooling summer breezes as well.

Design for climate involves three basic steps. First, the designer must become familiar with the macroclimatic conditions for the region in which the site is located. Second, the microclimatic influences that produce on-site variations must be determined. Third, the designer must decide from the program statement the critical relationships between outdoor and indoor areas and the climatic requirements of each area, and then must tie those findings to a study of the microclimate to determine the optimum location for each.

Compromise in designing for climate is almost always necessary, but the orientation of the structure and of the open spaces with respect to sun and prevailing winds should guide much of the detailed design. Once features are sited, design measures can be taken to provide more control, thus enhancing the livability of the site further. These measures include controlling wind through the use of windbreaks and chaneling, controlling solar radiation through sensitive choice of materials and the use of shade when necessary, and controlling precipitation through construction of overhead and vertical shelter and attention to drainage and low spots where precipitation might collect. Control of temperature follows from control of the wind and sun.

REFERENCES

1. Kevin Lynch, *Site Planning*, 2nd ed. (Cambridge, Massachusetts: M.I.T. Press, 1971), p. 68.
2. Gary O. Robinette, ed., *Landscape Planning for Energy Conservation* (Reston, Virginia: Environmental Design Press, 1977), p. 20
3. Victor Olgyay, *Design with Climate* (Princeton, New Jersey: Princeton University Press, 1973), p. 80.
4. Lynch, *Site Planning*, p. 65.
5. Olgyay, *Design with Climate*, p. 44
6. Robinette, *Landscape Planning for Energy Conservation*.
7. Olgyay, *Design with Climate*.
8. Olgyay, *Design with Climate*, p. 51.
9. Robinette, *Landscape Planning for Energy Conservation*, p. 11.
10. Olgyay, *Design with Climate*, p. 100.
11. Robinette, *Landscape Planning for Energy Conservation*.
12. Olgyay, *Design with Climate*, p. 45.
13. Olgyay, *Design with Climate*.
14. Robinette, *Landscape Planning for Energy Conservation*.
15. Olgyay, *Design with Climate*, p. 15.

4 Circulation

The circulation systems to, from, and within a site are essential to its use, and in many cases they can dictate the site's entire layout. Any type of movement through space is a form of circulation, whether it be by two or more wheels, by foot, by water, by rail, or by air. Circulation is necessary for living, working, playing, and engaging in simple conversation: people who mingle at parties are said to circulate through the crowd; blood circulates, as does water in a fountain. Circulation as the movement of people or of things needed by people through a site is the subject of this chapter.

In the earlier discussion of exterior space, circulation was described as the only way a person can fully experience the site in three dimensions. The constantly changing panorama of views and vistas of a site experienced through movement were described as more important than the view at any single, frozen moment. A variety of ways of experiencing a site (and a variety of approaches to it) can be created through changes in the circulation system. Circulation systems also fill a critical need in moving people from place to place and in servicing people with fresh information and goods.

TYPES OF SYSTEMS

There are basically three types of circulation systems that have distinctive influences on site, space, and structure. These are the pedestrian system, the two-wheeled nonmotorized system (involving primarily bicycles), and the motorized vehicles system (which includes everything from cars and buses to trains). Air, water, and rail travel, because of the highly specialized demands they impose, are not discussed in this book.

The Pedestrian System

The pedestrian system is characterized by looseness and flexibility of movement, slow speeds, and small, human scale. Of all circulation systems, it offers the most design freedom because it benefits from the human ability to climb steep grades, turn sharp corners, and change direction or stop abruptly. This flexibility of movement can create problems, however: since pedestrians essentially can go wherever they please, the designer must channel the flow to desirable locations. Too much rigidity in the design of the pedestrian system will be met with resistance. Too little control will lead to much of the site being trampled and abused by pedestrians in search of shorter routes to their destinations (figure 4-1).

A hierarchy of use intensities often develops on large sites when pedestrians are more or less allowed to choose their own direction: paths widen at heavy traffic locations, such as main arteries, entrances, and exits, and narrow at places of light traffic. This can be used as a guideline in the final design, with the paved routes ultimately corresponding at least in part to the highest-use areas.

Even the most carefully designed pedestrian network cannot be expected to keep people on the planned routes at all times. People's need for occasional speed will result in shortcuts, and changes in schedules may increase or diminish the numbers of people using a path at a given time.

The Bicycle System

The two-wheeled vehicle system is gaining acceptance in the United States, but has not yet reached the status it enjoys in Europe, where distances between destinations are much smaller and open spaces are more intimately detailed. Consequently, the large-scale circulation of bicycles, tricycles, and mopeds in the United States is often ignored or shuffled into the pedestrian or vehicular systems as an afterthought. Yet two-wheeled vehicles are not highly compatible with either people on foot or cars, and the bicycle system is characterized by speeds faster than walking speed but usually slower than driving speeds. Other special features of the bicycle system are need for storage at or near the final destination, seasonal use, ill-defined legal rules and regulations, and the ability of traffic to go nearly as many places as pedestrians can. In recent years bicycling has become a major recreational sport as well as a means of getting somewhere.

The Vehicular System

The vehicular system presents by far the most complex design requirements of any circulation system. This system is characterized by wide variations in speed and vehicle size, with corresponding needs for surfaced routes of different dimensions to provide adequate negotiating space in transit and adequate storage space upon arrival. Because of this system's size, technical requirements, and cost, its design frequently determines the layout of all other site elements. This is particularly true of sites affording limited opportunities for connecting the on-site system with off-site feeders, and of sites whose topographic conditions or existing facilities mandate a certain location. Often the issue of economy determines design along the most efficient route.

FACTORS INFLUENCING TRAVEL

Travel within any circulation system is either purposeful and destination-oriented (a matter of getting from point A to point B) or it is recreational (pursued for its own sake). Most systems must accommodate both uses, although they are usually designed for one or the other. Trouble can arise when recreational and destination-oriented travelers use the same system—a result of their differing expectations. Destination-oriented travel is more direct, and the users expect that this system will be faster,

4-1. The pedestrian circulation system.

whereas recreational users expect to travel through reasonably attractive settings at their leisure. Combining destination-oriented travel with recreational travel can be dangerous, can reduce efficiency, and can result in two disgruntled groups of users.

For practical purposes, no circulation system exists independently of at least one of the others. Even systems used exclusively for recreation must somehow be reached by prospective users. Consequently, different systems meet, cross, run parallel to one another, and diverge.

The points of transfer between systems are critical in terms of the overall function of the site. Transfer points may be internal—with a significant change in direction or speed within a system, such as occurs when moving from a residential area onto the approach ramp of a freeway—or external—as when the driver of a car parks it and becomes a pedestrian to get into the building, or when a bicycle rider reaches a central storage area, stores the bicycle, and either boards a bus or begins walking. There are also subcategories of internal transfer: an automobile owner

4-2. Transfer between or within systems may involve a complicated interchange. Photo courtesy Soil Conservation Service.

may drive to a parking location and board a train or bus that uses the same circulation corridor as automobiles use, yet the transporting vehicle is different (figure 4-2).

A circulation system network combining pedestrian and vehicular traffic is only as efficient and as fast as its worst, slowest intersection or transfer point. Six smoothly flowing lanes of traffic, that are squeezed down to only two lanes for a short distance will cause the whole six-lane system to function for several miles in back of the squeeze point as though it were only a two-lane system. Similarly, only so many pedestrians can stand on a street corner before they begin getting in the way of oncoming traffic.

Origin-Destination

The layout of the circulation system must be functional, getting people to their desired locations in a safe, efficient, and pleasant manner. The use of origin-destination studies can help fix the locations of the travel corridors. Origin-destination studies consist of straight lines between points of ingress or egress, often represented in varying widths or intensities to denote greater or lesser use.

From a purely functional standpoint, the ideal placement of walks, drives, and other circulation corridors would match the alignment indicated by the origin-destination studies. Using this as the only design criterion, however, does not take into account aesthetics, technical aspects of the site (such as the presence of a steep grade or of features requiring preservation), or slight shifts in use patterns that can change the locations of the paths. Nonetheless, using an origin-destination study as a guide, the designer can compromise, manipulate, and maneuver the users to follow a system that provides a nearly direct route between locations, yet considers aesthetics and economics as well (figure 4-3).

Topography

The many variables of a site's character also influence the design of its circulation systems. Topography can limit the designer's choice of locations to one or two possibilities because of the high cost and logistical impracticality involved in drastically altering the grade. This is especially true for vehicular systems. The slopes of a circulation system must fall in a relatively narrow range between excessive steepness and flatness if the system is to be usable.

There are two basic ways of handling steep slopes in the design of its circulation systems. Topography can limit the designer's choice of locations to one or two possibilities because of the high cost and logistical impracticality involved in drastically the grade. This is especially true for vehicular systems. The slopes of a circulation system must fall in a relatively narrow range between excessive steepness and flatness if the system is to be usable.

this approach, and the landforms retain much of their original appearance; however, the amount of surfacing required increases significantly, pushing up costs of materials. Aligning the

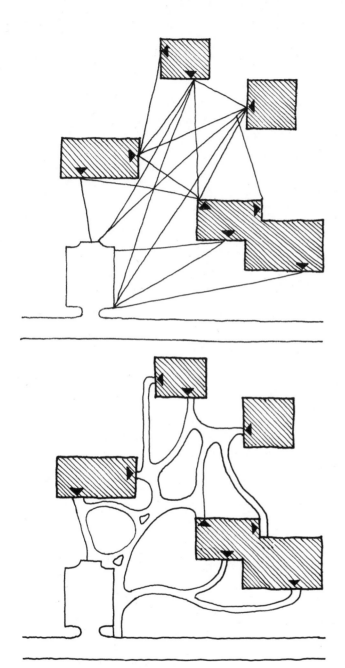

circulation corridors at an oblique angle to the topography is not desirable from an aesthetic standpoint, although it may work technically (figure 4-4).

Climate and Vehicular Travel
Sun and Snow

The climate of the area, (specifically the microclimate of the site itself) influences the design of the travel corridors by creating areas of sun and shade, or by raising potential problems of ice buildup or runoff. The sun becomes a factor in the alignment of east-west corridors that are expected to receive high use at sunrise or sunset. Such an alignment can severely impair a driver's vision during certain times of the year, causing a safety hazard. Circulation systems located on north-facing slopes in the cool and temperate macroclimatic zones may have problems with ice and snow buildup, causing hazardous conditions for vehicles and pedestrians alike. Unless properly designed for drainage, low spots in the corridors can become ponds, icy spots, or gathering places for fog, all of which are dangerous and increase maintenance costs.

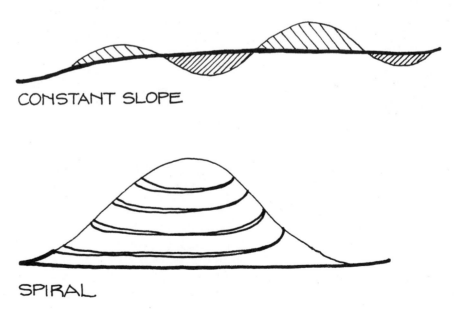

4-3. Origin-destination studies can help determine the most functional routes for users.

4-4. Two methods of dealing with steep topography: a. a constant slope; b. spiral.

Wind

The positioning of travel corridors in relation to prevailing winds should also be considered. Turbulent, gusty winds blowing through draws can create serious problems for high-profile vehicles such as campers and can force bicycles into the path of oncoming traffic. East-west roads are likely to have sections that blow free of snow in the winter—which is ideal except in blizzard conditions, when the sheets of snow can cause zero visibility and can cause certain sections of the roadway to drift completely shut. Although the drifting and blowing can be controlled to an extent through careful planning of roadside cuts and fills and through use of snow fences or windbreaks, complete control is not possible. On the other hand, alignment of critical travel corridors directly into the winter winds willl allow them to blow clear—albeit directly into the face of the windward-bound traffic.

Climate and Pedestrian Travel

A slightly different set of climatic conditions must be considered in the design of the pedestrian circulation system. For example, glaring sunlight can more seriously incapacitate drivers turning through intersections than pedestrians waiting to cross. An essential goal in the design of the pedestrian circulation system is to create a comfortable environment in which to walk. Pedestrians who have to walk a long distance will appreciate a sidewalk that is shaded in the summer and open to sunshine during the winter, that does not force them to face directly into strong winds, and that does not present abrupt changes in grade inviting a fall or taxing out-of-shape people beyond their limits.

The hazards of poor drainage and faulty orientation that result in ice problems for vehicles are even more pronounced for pedestrians. Because of age, infirmity, or handicap, many persons must use a wheelchair or cane to get around, and for them, the site is rendered unusable by patches of ice, rough or broken surfacing, or extremes of topography. Much damage and loss of life connected with weather-related vehicular accidents is significant, but pedestrians are the ones who suffer the shattered ankles and elbows, the broken hips and collarbones from tumbles on ice or rough pavement.

Climate and Bicycle Travel

Bicycles and other nonmotorized vehicles are limited in use to the warmer seasons of the year, although sporadic use occurs during the winter months by a few people. The braking ability of bicycles (severely reduced on ice and on rough gravel or sand) determines the surfaces over which bicycles can negotiate. The lightweight construction of bicycles makes them unsuitable for use in heavy snow or water conditions.

A close relationship exists between steepness of grade and sharpness of turn: a bike rider cannot be expected to negotiate a corner that has a five-foot radius at the bottom of a 10% or 15% grade when his or her speed has built up to thirty-five miles per hour. The turning distance required increases with the speed and angle of approach.

Adjacent Systems
Site Accessibility

The adjacent circulation systems often determine what systems can be used on the site itself. In fact, ease of access for the type of vehicle required is often an overriding reason for selecting a site in the first place. If a client must have rail service for the transport of heavy goods, or receives shipments by semi- or double-trailer truck, the most economical sites will be those that are readily accessible to existing rail or trucking lines. Lack of such availability will mean that the client must rethink the methods of delivery, or must build the access lines or trunk lines as a part of the project, or must move to a different area.

The designer must analyze how and when the users of a site under development will be entering and leaving the site. This analysis can be based in part on how the majority of users presently travel to and from the existing site—what streets and highways they use, where bottlenecks occur, whether any alternate means of transportation is available. If the designer concludes that the daily commuter will continue to arrive at the new site by private automobile, the site should be located and its access points designed to allow quick, easy ingress and egress for cars. If great numbers of users arrive by public transportation (either buses or subways), the distance from the street to the major site entrances and particularly to covered walkways or other protected paths, for use in inclement weather, should be made as short as possible.

The above points relate not only to the adjacency of systems that might be used on the site, but also to how well the site functions overall; no matter how beautifully a site is designed in terms of aesthetic quality and spatial relationships, if it cannot be reached, it will not be used. Compromise must usually be made between the most interesting circulation systems from a design standpoint and the most efficient systems from a travel time and cost standpoint—keeping in mind that people traveling to a destination usually show positive interest in what occurs along the way only if such occurrences do not interfere with their travel time.

A system that deviates slightly from the straight and true will remain functional, but one that sends people in what they perceive as being the direction opposite that in which their destination lies will not be used. Vehicular drivers, of course, have little choice, and (since many sites are so small that the only vehicular system needed is one to get from the adjoining street into the parking lot) as long as the curves and grades are negotiable and the users are aware of their destination, the system will function. Pedestrians, however, are much more likely to strike out on their own if the system does not function according to their needs (figure 4-5).

Aesthetics

Aesthetics is another important consideration in the design of circulation systems. Any drive or walk can be made less monotonous and more interesting through design attention to the route's alignment, views, and vistas, to what occurs along the sides of the route and ahead of it, to how the site is perceived as one travels along one of the designated routes, and to whether or not the route is clearly understandable. Designers must avoid the twin pitfalls of a system that is so interesting that it becomes unsafe because the attention of the users is constantly directed away from traffic and a system that is so boring that the users' minds wander in an effort to create their own interest.

All of the theories used in the design of exterior space—color, balance, form, line, texture, rhythm—come into play to create interest in the circulation system, which is a part of that exterior space. The aesthetic appeal of the system will largely determine the mood of the users upon arrival at their destinations. If the travel corridor is functional, pleasant, safe, interesting without

being overwhelming, and directional without being overpowering, the users are more likely to arrive in a state of mind conducive to work or relaxation. A drive through depressing, cluttered, dirty streets clogged with traffic that backs up at every light, with nothing for the drivers to look at as they wait, will tend to create a tension that grows progressively worse with each delay.

For pedestrians, a walk sheltered from the elements and scaled to the human figure instead of to the freeway is much more likely to evoke a positive response than one whose users are forced to

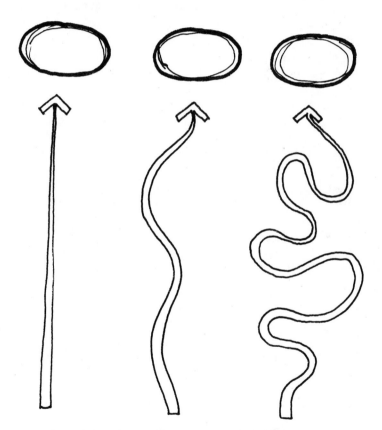

4-5. Slight deviations from a direct route are desirable in most circulation systems, but too much deviation will result in decreased use.

dodge potholes, step over trash, and wonder what new menaces lie around the corner of the next building or behind the next row of shrubs. Bicycle riders will find that a route free from conflict with pedestrians and vehicles, well-lighted at night, and sloped to challenge without defeating is more attractive than one on which they have to fight with vehicles for space or worry about running down pedestrians at every corner.

Quality

People will use an established circulation system if they perceive it to be safe, functional, efficient, and headed in the direction they need to go. The quality, speed, and alignment of the system all contribute to the users' perception of its usefulness. Quality relates to both the basic considerations of maintenance and the aesthetics of the system. A travel corridor that is kept free of snow, gravel, or debris and is repaired each season will appear safe and comfortable. How well-directed the system is provides a measure of its quality as well. Many college and other institutional campuses are notorious for their lack of signage or for signage that seems intended only to direct people who are already familiar with the layout.

Speed

A circulation system should be designed to operate with maximum efficiency at the speed at which the users will most often travel. A high-speed, efficient, get-there-from-here system, such as the interstate highways, is of necessity straighter and less hilly than a system designed to be used by local traffic only; its transitions must be laid out gradually, whether into curves or into on and off access corridors. The alignment, gradient, type of paving used, number and location of access points, and design and location of focal points and other attention-getters should all be considered in terms of the speed of the system. A system planned for high speed but detailed with sharp curves and steep grades cannot be traveled at the design-optimum speed—reducing the system's efficiency and the satisfaction of the people using it, and perhaps even forcing people to look elsewhere for a more useful corridor. The type of paving used, its roughness and variation according to weather conditions, and the degree of maintenance it receives contribute to the speeds users can attain while still feeling safe. A shifting pavement such as gravel becomes unsafe at high speeds around curves and over hills (as well as during snowy or wet weather), thus reducing the safe maximum speed.

Control of Access Points

The most efficient, fastest travel system will become a nightmare to use if the access points are not adequately controlled. The more intersections there are and the closer together they are located, the greater the potential for accidents becomes and the slower the safe speed must be. Finally, the details used to attract and keep the user's attention must be scaled to the speed at which the system is intended to be used.

Use of small, finely textured or finely scaled details close to the drivers on an interstate system will distract them from the road ahead, causing the drivers either to slow down or to quit paying adequate attention to what is going on in the road. The details and attention-getters for a high-speed system should be located within the cone of vision naturally maintained by drivers going in the correct direction, rather than being located outside that cone and requiring a potentially deadly shift of attention; in addition, they should be scaled to be legible from a long distance, thus not forcing a sudden change of visual alignment that can communicate itself to the wheels of the vehicle. Instead of individual tree or plant specimens, the designer should use masses, and instead of small, finely lettered signs with a lot of information, the designer should arrange for signs that give the basics only and in large enough letters to be absorbed in one glance (figure 4-6).

The detailing for the pedestrian system should be fine and should attract attention close at hand. In this system a single plant or a change in paving material becomes important. The use of too-large materials, outsized masses, and coarse textures intended to be seen at a distance will only make the pedestrian feel out of scale with the environment and will make travel along that corridor uncomfortable (figure 4-7).

The bicycle-moped system falls somewhere between the pedestrian and the vehicular systems. While a bicycle can reach speeds in excess of forty miles per hour, the cyclist is nevertheless more exposed than a driver is to the road and to the details at the side of the road; the cyclist can also stop more quickly and turn more sharply than a driver can. The signage and details for cyclists should relate to the scale of their vehicles, particularly when signaling a stop or other directional or speed change.

Matching Scale to Speed

Serious problems occur when the scale of the design does not match the speed of use. A good example is the interstate system, which was designed for speeds of seventy or more miles per hour. With the reduction in the speed limit to fifty-five miles per hour due to the energy crisis, the system has become monotonous—making it more difficult for drivers to keep their attention on the road (or to resist speeding).

a.

b.

4-6. Detail is an important part of circulation design: a. fine details distract drivers at high speeds; b. masses at distances are more appropriate.

4-7. The pedestrian system should be finely detailed, not overwhelmed by coarseness.

Alignment

Alignment relates not only to straight path versus curvilinear route, but to the number of hills and valleys, their steepness, whether the curves occur at the top of the hill or at the bottom, how clear the sight distances ahead are, and how much visual interference occurs from off-site factors. The straighter and smoother the route's alignment is, the higher the safe maximum speed will be, and the faster the system as a whole is likely to be. The hillier the system, the more caution must be exercised, in case something unseen is blocking the road ahead, or in case an

abrupt change in direction with limited adjustment distance lies just over the crest of a hill. When such a change in alignment does occur, the users may not be aware of the change until they are nearly into the curve, at which time they may be forced to brake quickly, causing their vehicles to slide.

Sight distance can easily be blocked at a smaller scale in urban areas when inadequate distance is allowed between intersecting streets and the corner of a building or a tree mass. Fortunately, sight distance guidelines now exist to control the height and setback of any new construction that might interfere with a clear view of an intersection. In older developments, however, the hazards still exist. This type of sight distance interference is particularly dangerous to pedestrians and cyclists because a vehicle will often pull into the center of an intersection to get a clear view before proceeding through it and may not see an approaching cyclist or a pedestrian who just stepped off the curb. A designer who intends for travel to occur at slower speeds can make use of sharper, less clear alignments to do so, if proper safety precautions are taken (figure 4-8).

DIRECTING MOVEMENT

Once the designer has given due consideration to questions of the primary type of vehicle to be accommodated, optimum speed, desired alignment and level of quality of the system, the more subtle design problem of actually directing movement along that system can be undertaken. A circulation system can be designed either as a direct route (obvious from start to finish) or as an indirect route (involving decision points where the route and direction of the destination are not quite so obvious). Either type still requires reinforcement of that system through sensitive design of all other aspects of the site, in order to make the system interesting and to make the spaces associated with the system and the destination appear connected to one another.

Visual Reinforcement through Vista and View
Vista

The use of visual reinforcers of the circulation corridors is one way to assure that the system will fit in with its surroundings, rather than imposing on the site. Visual reinforcement occurs through the use of the design elements of view and vista, correlating these with changes in direction or directional site features.

4-8. A tortuous alignment reduces efficiency and safety.

In direct routing, alignment of the route along a vista to a terminal space or feature is one way of holding the interest of the traveler (figure 4-9). This alignment allows the terminating feature to be revealed in progressively greater detail. Difficulty may arise with such a direct approach, however, if the destination happens not to be the visual terminus: focusing on the visual terminus in such a case can cause the users to miss their turn.

The distance between viewing station and terminus can also have the negative effect of making the actual travel time seem longer. From the viewpoint of the pedestrian user, the slower speed of travel on foot and the more detailed association with the site work together to make the path to the vista terminus more compelling; but the lengthening of the apparent distance, too, can be more strongly felt at the pedestrian scale (figure 4-10). The designer might combat this by breaking the travel experience into a number of smaller sequences, each building toward the final terminating feature but having interest and merit in its own right.

When the vista is used as a reinforcing visual part of the travel experience, the travel corridor acts like an axis, particularly since most circulation systems are two-directional and must function for individuals whether they are coming or going. If two-way travel is not taken into account, travel in one of the directions may have an inadequate visual focus or termination.

The vista (which can be part of the overall view) can also be used to reinforce travel along an indirect route, by being structured toward a focal point that occurs at a change in direction. This helps to break the monotony of a continuous landscape that changes little along the sides of the corridor. A vista can become so repetitive as a reinforcing element, though, that it loses its meaning—the movement becoming static and the terminus of the vista losing its ability to attract and direct motion.

4-9. A vista can hold attention.

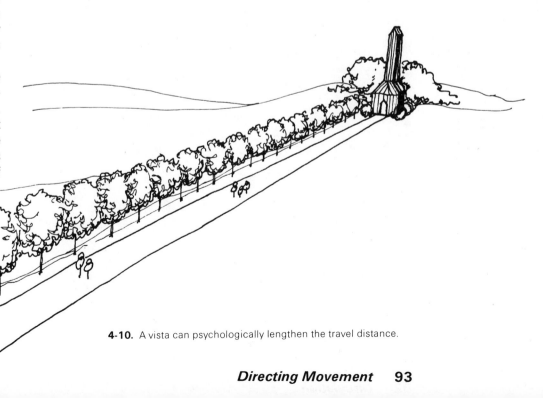

4-10. A vista can psychologically lengthen the travel distance.

View

The use of a view to reinforce motion is often more casual than the design of a vista. One type of view is panoramic, with travelers winding their way through it, getting ever closer to a destination. This can be experienced on a road from the mountains to a city located on the plains: the city is visible for a great distance, but the view of the city constantly changes as the traveler descends through space and becomes part of the scene that was previously experienced only visually.

Alternatively, the view may be calculated to attract attention to and build anticipation for what lies ahead. In this case there may be a glimpse of an interesting space or portion of a structure, through a break in the trees or around a bend in the road, and then the view may be directed straight ahead for a distance, only to offer another, slightly different glimpse around another bend or over another hill. The psychological experience for the user is quite different in each of these examples.

4-11. A winding approach sensitively keyed to the experience.

Total visibility of the destination—or at least of what appears to be a space or structure at the end of a journey—can build feelings of comfort and relief (particularly if the travel corridor is descending) or of awe and wonder. A winding approach to an imposing monument or mansion lends itself to the latter type of feeling, even though the traveler is aware of the structure for the entire travel time.

On the other hand, the concealing and revealing of the spaces and structures makes the traveler wonder what to expect. The indirect approach must not be too indirect, nor the destination too concealed, or the user will become frustrated. This is particularly true of the pedestrian system, where an error in direction takes much more effort to correct (figure 4-11).[2]

Spatial Transitions

Another way to reinforce circulation corridors is to create smooth, interesting spatial transitions from one location to another. The sequence of experience is more important to a system than the single experience in a given spot. The transitions in this sequence may reflect different intensities of use, with the detailing of the system changing to correspond with these differences. It may consist of transitions of materials to match changes in the countryside from natural landscape to suburban to urban to inner city.

Unfortunately, while many people travel through this last type of spatial transition daily, it is usually poorly thought out in relation to the overall circulation system. Instead of creating a series of connected experiences, with the transition of materials gradually becoming more urban and less rural, the transitions commonly take the following form: the user is spiraled off the interstate (with its large-scale detailing and open spaces) onto a ramp that curls and winds between utility poles and signs, and then onto a spur highway or urban sprawl corridor that often features nothing but guardrails and barren industrial buildings, and from there into a densely planted residential area. No transition is provided between spaces along the travel corridor to correspond to the spaces connected by the corridor or to the spaces between the spaces themselves; and as a result, the travel experience lacks continuity.

Transitions may be handled in terms of relative openness and enclosure. A site intended to be inviting to the public may present a corridor that moves through ever-widening open spaces, progressively better-lighted and less enclosed, until the final space is reached. The opposite effect—that of coming to a place of isolation or privacy—can be reinforced by gradually narrowing the path, creating tighter and tighter enclosure, until finally the destination is seen as only a small break in the landscape.

Negative Reinforcement of Movement

Movement along circulation corridors can be negatively reinforced by blocking passage to certain places or along certain routes, as opposed to encouraging travel. Methods of blocking may resemble those used in encouraging travel—the use of enclosure or of view or vista—except that the connotations of the devices presented are negative: cul-de-sac streets, "T" intersections, and wrong-way-do-not-enter signs are examples.

For the pedestrian, lack of a paved surface may not be enough of a deterrent, and the designated path may require reinforcement by a barrier at the decision point, such as a hedge, a fence, or a wall. A low quality system—incorrectly designed for speed, awkwardly aligned, and dependent upon indirect negative reinforcement—will keep people away rather than encourage use.[3]

Sensory Reinforcement

Reinforcement of movement occurs not only visually but also as the other senses respond to the space. Given a choice of two paths, one past a bakery and the other past an oil refinery, a pedestrian may almost unconsciously seek the bakery route. What people hear can also direct motion along a given path. The sound of a band playing in the distance or of water heard but not seen can draw people into and through a space. The sound of an oncoming train, almost as much felt as heard by the hum of the tracks, can lead people either to traverse the intersection quickly, or to hurry off in another direction to avoid the conflict.

The success of a route (as well as the effectiveness of its subtler reinforcers) is also dependent on the climate, time of day, user's mood, amount of time available for travel, and many other factors over which the designer has little control. The designer does well to recognize these limitations and to accommodate the users' free choice.

CIRCULATION AS SEQUENCE

The design of circulation systems does not occur independently of the design of the spaces and structures, but integrally with them. The entire spatial experience is one of sequence, whether the user is traveling to a site, through a site, or down the corridor of a building.

The sequencing of movement can either encourage travel to continue in a similar manner (with approximately the same direction and speed) or cause people to slow down, stop, or speed up. This depends upon the handling of the reinforcing elements that occur along the travel corridors: the sequencing of open and enclosed spaces, the positioning of views and vistas, where the balance occurs, how color is used. Setting up a rhythm that is unbroken will encourage continued movement. Breaks in the rhythm will cause users to slow down or stop.

For the pedestrian flow, the calculated use of stopping places can increase use of shops and restaurants or can control the backup of people trying to cross a street. The pedestrian circulation system can be a richly varied set of experiences, but many so-called pedestrian malls fail, for two reasons: first, they are scaled beyond the pedestrian's comprehension; and second, they are too monotonous, with each space too similar to all the rest.

DESIGN OF CIRCULATION SYSTEMS

Designing a circulation system as part of a site's exterior space can be approached in two ways: the system can serve as a spine, with basically unrelated activities and experiences taking place along it; or the system can be nodal, with the circulation system and the spaces essentially interconnected. (figure 4-12).

Circulation Spine

The circulation spine is used in many street systems, where the overall circulation is continuous, but each space or structure is a destination on its own. The distinction between the street or path and the spaces in such a system is obvious, and once people leave the travel corridor, they have little reason to return to it again until they are ready to leave the space for another destination.

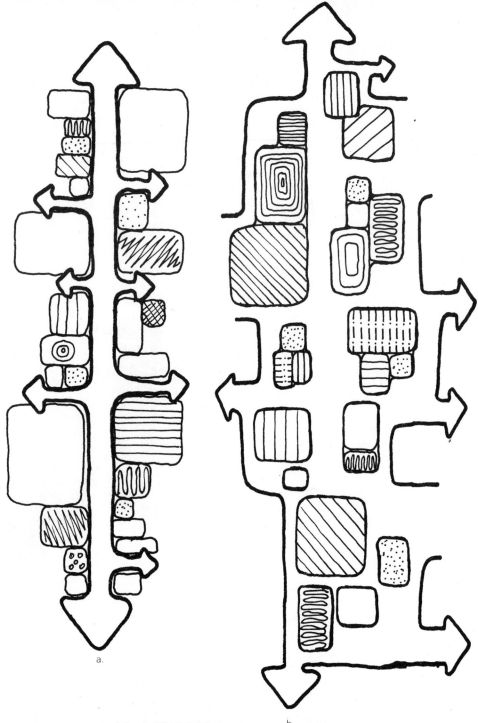

Circulation Nodes

The distinction between street or path and surrounding spaces is much less clear in nodal circulation systems. In such a system, the street may widen to include a small, screened parking area, or the sidewalk may become a route through a seating area. A plaza or greenspace may blend almost unnoticeably with street-side planting. The street may narrow and the pedestrian flow traverse it with no break in stride—the spaces working back and forth rather than assuming the linear form suggested by the edges of the pavement or circulation route. The fabric of the streets and the spaces, whether for pedestrians or vehicles, is interwoven.

Nodal circulation systems make for a varied, interesting experience. However, they can also be dangerous, unless designed with an understanding of which system (vehicular or pedestrian) will dominate and which will be subordinate. Examples of the successful use of this type of system are the Country Club Plaza in Kansas City, many European cities, and some institutional campuses.

Suggested Movement

Much of the discussion up to this point has assumed the system's use of a paved, obvious path for travel. However, many of the most delightful travel corridors for pedestrians and bicyclists are not much more than a path defined by plant materials or fenceposts. The designer can use enclosure and visual quality to direct a pedestrian between a break in the trees or along a path that leads to an open space, without having to define that path by using stepping stones or pavement. As long as the route is supported by the visual and physical experiences along the sides, it will be a route that is used. This relates back to the first step in designing space: designing with forms rather than being dictated to by details.

Technical Organization of Circulation Systems

Circulation systems can be organized in any of a number of general patterns, depending on economy, the direction and carrying capacity needed, site conditions (including topography and climate), and the vehicle used to negotiate the circulation system.

4-12. Circulation: a. spinal; b. nodal.

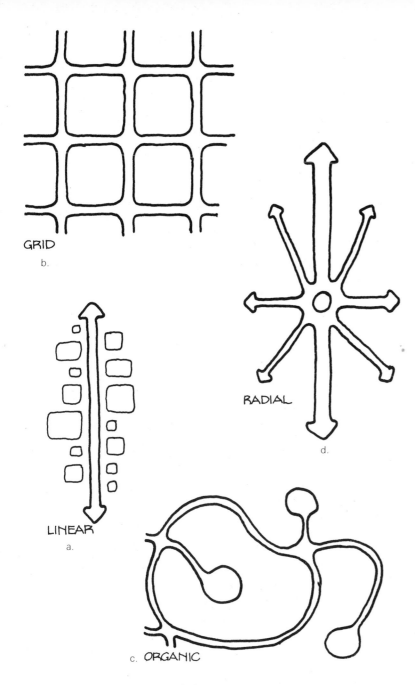

GRID

b.

RADIAL

d.

LINEAR

a.

c. ORGANIC

4-13. Four ways of organizing circulation systems: a. linear; b. grid; c. organic; d. radial.

Four general patterns form the basis for many of the individual variations in circulation systems. These are known as the linear, grid, radial, and organic (or disordered) systems (figure 4-13).

The Linear System

The linear system is characterized by continual lines of movement in one or more directions. The system can become quite congested if it is designed for unlimited access—particularly if vehicles are allowed to back out of drives and parking spaces directly onto the driving surfaces. On the other hand, the system can be inconvenient to use if access is very limited.

Often, roads leading into cities provide clear examples of linear circulation systems: straight, fast-moving, double- or triple-lane routes paralleled by a frontage road that allows access from private property onto the thoroughfare. Unless controlled, linear development can go on and on, eating up acres of land in a pattern that is awkward to expand upon. This type of system is more commonly spinelike than nodal (figure 4-14).

4-14. The linear system.

The Grid System

A prime characteristic of the grid system is that it allows free movement in many different directions. The common division of urban areas into city blocks and the plotting of rural road systems composed of a road every square mile are examples of grid circulation systems. The farther apart the intersections are set, the more smoothly traffic flows—but concurrently the less convenient the system is in terms of free movement.

The grid system is often superimposed on a landscape without regard for natural conditions or for existing features that could enhance the character of the area if preserved. Variations on the standard rectilinear grid do exist, however, and these pay more attention to site conditions by curving and varying the distance between intersections in response to traffic patterns and to natural features as well. The grid system may combine spinal and nodal development, depending on the sensitivity with which the adjoining spaces are designed (figure 4-15).

The Radial System

Radial circulation systems involve the convergence of traffic at a central point, which is functional and convenient as long as the point is the destination of the travelers. The radial system is dominant, structured, and usually formal; it lends itself well to supporting important monuments and major central spaces, such as city or town squares. In terms of economic development, the radial circulation system produces many odd-shaped, hard-to-sell triangles of land. It also poses difficulties to smooth crossover into a grid or other system.

The Organic System

The organic circulation system is the most sensitive to the conditions of the site—sometimes at the expense of logical functioning by the system and of easy interpretation of it by the users. The dead-end or cul-de-sac street, curving and recurving paths, and abrupt changes in direction all characterize the organic system.

Use of this system often involves relatively little disturbance of the soil, topography, and existing plant materials. People may experience difficulty, however, in finding their way around an organic system: streets may change names at changes of direction rather than at defined intersections, and a street of one

4-15. The grid system.

4-16. The organic system can be difficult to follow.

name will often end in a "T" intersection, only to pick up with the same name several blocks away (figure 4-16).[4]

The pedestrian system is usually a combination of one or more of these general circulation patterns. Because of the greater flexibility of movement in the pedestrian system and its less demanding technical requirements, the system can change direction quickly to respond to off-site influences or features that should remain undisturbed. Pedestrians will rarely follow a geometrically perfect line, particularly if it involves turning a 90° corner, unless they are forced to do so by conditions that keep them on the path. Therefore, a pedestrian system designed around order and symmetry might include angled walks at the intersections rather than 90° corners.

DESIGNING CIRCULATION FOR CLIMATE

Certain design concerns must be addressed regardless of the general circulation pattern chosen for the site. Foremost among these is designing the system to solve existing climatic problems and to avoid creating new ones.

Since the climate of a site influences its comfort, which in turn influences the use it will receive, circulation systems should be designed to provide maximum comfort for their users, whether walking, bicycle-riding, or driving. Many of the problems associated with climate can be solved with appropriate detailing. For example, problems with ice buildup in low spots or with the gathering of fog that reduces sight distance can be avoided by eliminating low spots when designing the road profile. The use of snow fences along the sides of roads or walks will keep drifts from encroaching on the travel corridors, reducing hazards and the amount of labor needed to keep the systems open. Windbreaks in combination with snow fencing will keep swirling snow to a minimum.

When possible, driving surfaces should be located south of shadow-producing elements, such as structures, to allow them to warm up and rid themselves of ice during the winter. This is even more important for the pedestrian system, where reduction of ice and snow hazards during the winter and reduction of heat and wind during the summer (by orienting the walks to take advantage of shade) will add to the pleasure of the experience.

THE PSYCHOLOGY OF CIRCULATION DESIGN

The psychology of designing circulation systems was discussed. In addition to relating the scale of the details used to human size, the designer can affect the psychological impression conveyed by a circulation system by reducing or increasing apparent distance. The visual focal point of a system can be adjusted into the foreground or background, depending on whether it is meant to serve as an immediate reference point or as a distant one. The physical alignment of the corridor can twist and turn, or it can be direct and straight. Either of these designs can feel long and drawn-out: the twisted one because the drivers or hikers feel as though they are looping back and forth without going forward; the straight one because no visual change breaks the monotony along the way.

A twisted path can be helped by adding directive elements (to reinforce the users' sense that they are going in the right direction) and detail elements (to enhance interest enough to keep the users' minds off the circuitous route). The straight path can be made more interesting by interrupting the simple, direct line with cross streets or intersecting paths, or by varying the density of the enclosure used to reinforce the movement. The amount of change needed to keep the system interesting and functional depends on the system's scale and speed of movement: too-frequent interruptions in the vehicular system will be confusing and dangerous, fragmenting the attention of the driver; too few in the pedestrian system will make each block seem longer than it really is.

CIRCULATION SYSTEMS DESIGN PROCESS

The first requirement in designing circulation systems is familiarity with the site and program conditions, gained through a careful analysis of the site. This analysis should take into account the parameters that cannot be changed, as these will determine the type of system used and probably its location. Topography, vehicular movement, adjacent systems, and cost can all be fixed parameters in a project.

The next step is schematic site layout—preferably a concurrent layout of site areas and circulation so that the designer avoids imposing an incompatible circulation system on the site.

The functional requirements of the users come into play here, with origin-destination studies or straight-path layouts being used to place the corridors where they are most needed. This step is followed by refinement of alignment, and by compromise between the optimum circulation corridor and the optimum designs of the other site areas. Finally, the system is detailed for construction, with careful attention to alignment, sight distances, materials, and intersections.

DETAILING CIRCULATION SYSTEMS
Details and Safety Control

The detailing of circulation systems can make the difference between their being used or standing idle. Since safety is a prime consideration, the design of the places where systems parallel or cross one another should be given special attention. Changes in

4-17. Horizontal separation of systems, with vertical definition provided by light poles and trees.

direction and changes from one system to or through another can be safely handled in many ways, including by controlling the intersections through separation in space or separation in time.[5]

Spatial Separation of Systems

Separation of circulation systems in space can be accomplished either vertically or horizontally. Forms of vertical separation include skywalks and pedestrian overpasses (over highways) and ramps and elevated expressways (over slower-moving streets or rail lines). Vehicular use of vertical separation is so common that it is almost taken for granted. However, channeling pedestrians up and over a street rather than letting them dodge between cars on grade is less customary. As a result, unless the approaches to a pedestrian overpass are controlled to force its use, or unless the design is subtle enough to make it appear as direct and convenient as the more dangerous on-grade route, the overpass will not be used.

The underpass is another method of vertically separating circulation systems. Construction of an underpass or tunnel often involves a considerably greater disturbance of the surrounding grade than does an overpass, and the aesthetics needed to make the systems appealing to the users are more difficult to deal with. Since tunnels are usually long, narrow, and dark, their use, especially by pedestrians, must often be encouraged by the negative reinforcement of the on-grade routes.

Horizontal separation of systems involves their adjacent placement. Usually, the greater the separation is, the safer each system will be. New bikeways are often placed in the parking lane of an existing road, separated from the vehicles by small bollards or (worse) by a painted line, and the pedestrian sidewalk may be located just against the curb. The hazards of such designs are increased severalfold at intersections, where bicycles in the far right-hand lane must cross all lanes of traffic (from all directions) when turning left.

Nuisances also occur with this type of design—such as pedestrians and bicycle riders being splashed by automobiles on rainy days. Buses present still another problem, since they must pull over to the curb to allow passengers to get on and off. When this happens, bike riders must either wait behind, in the exhaust fumes, or pull into the driving lane to go around. A successful solution to the bus problem, implemented in several cities, is for the bus to use its own center or curbside lane (figure 4-17).

The greater the horizontal distance between systems, the safer they will be. However, assuming that bicycles or feet are frequently used as substitutes for motorized vehicles, the origin and destination of users of different forms of transportation will be the same, in which case the systems are likely to parallel or cross one another.

Time Separation of Systems

Separation in time can be as sophisticated as closing a system to all but one type of travel during certain times of the day, or it can be as simple as installing a painted crosswalk and a "Yield to Pedestrians" sign. The most common arrangement of time separation is by means of controlled signals at crossing points, stopping traffic in one pair of directions and allowing pedestrians and vehicles to move in the other.

The degree of control required in such an arrangement depends upon the speed of travel and the intensity of use. On low-use residential streets, a painted crosswalk may be all that is necessary. To allow pedestrians to cross six or eight lanes of traffic, however, a safe landing spot at the median may be required, as well as a long "Walk" signal during which traffic stops. Either the pedestrian or the vehicle may be given dominance, depending on whether the walker is allowed to move immediately when the light turns green or has to wait until turning traffic is through the intersection. At some intersections, vehicular traffic is stopped in all directions, allowing pedestrians to cross on the diagonal as well as along the side of the intersection.

The introduction of legal right or left turns on the red light has decreased the safety factor of the controlled crosswalk; more accidents occur because of a vehicle hitting a pedestrian when turning right through an intersection than at any other point. Even the most advanced control mechanisms operate to no avail if unclear sight distances are provided for the pedestrian to see an oncoming car or for the driver to make sure the intersection is clear of bikes and people before proceeding through it.

Alignment Techniques

In the alignment of circulation systems, two dimensions must be considered: the horizontal alignment (the placement of the curved and straight sections in relation to the horizontal ground plane) and the vertical alignment (the relationship between the slope and the horizontal alignment). Failure to consider the two dimensions in relation to one another can result in a curve occurring at the crest of a hill where sight distance is limited, or a sag occurring at the bottom of a hill where water will collect, or sharp transitions occurring between slopes and flat areas that can cause damage to vehicles.

For complicated circulation systems where accuracy is critical, the curve in both planes should be calculated and engineered. In all cases, the design of the alignment begins with a rough, freehand layout in plan and profile of the desired curves and patterns. For the pedestrian and bicycle systems, a slight refinement of this first design may be all that is required—particularly if the system is characterized by little slope or by gradual, smooth transitions between the slopes of the existing grades. Park roads and other lightly traveled roads may also be designed in this simple way. Most vehicular systems, however, must be engineered in order to attain the standards of construction needed to allow their intended vehicular use.

Horizontal Alignment

Horizontal alignment is based on straight tangent lines joining circular curves at tangent points. Spiral curves are sometimes also used in horizontal alignment, but applying them successfully is more difficult.

Several types of curves can be used in laying out a system. The reverse or "S" curve is an aesthetically pleasing, easily negotiable curve if the two curves are joined by a tangent section that gives the user a chance to relax before turning again.

The spiral is used occasionally to ascend or descend steep slopes, or to leave a fast-moving system such as the interstate and enter a slower-moving feeder system. It allows the user to increase or decrease speed gradually, and it can compensate for slopes that would be far too steep to negotiate by a path going up and over the hill. Use of a spiral curve must take into account the amount of work needed to steer the vehicle around the curve: if a driver must constantly adjust his or her hand position on the wheel, more attention is needed, and the possibility of accidents increases.

The broken-back curve combines a radius in one direction with a straight line joining a radius in the reverse direction. This is an uncomfortable curve to negotiate, especially if the straight-

"S" OR REVERSE

REVERSE WITH TANGENT

SPIRAL

BROKEN-BACK

4-18. Different curves used in horizontal alignment.

line tangent section is not very long. It can also be a difficult design to resolve aesthetically.[6]

Most sites exhibit many different types of curves (figure 4-18).

The following components are involved in the calculation of horizontal curves (figure 4-19).

Tangent (T): A straight-line section of the road or walk, measured from either the point of tangency to the point of intersection, or from the point of intersection to the point of curvature. The two tangent sections of a curve measured in this way are equal.

Point of Curvature (P.C.): The beginning of the curve, where the tangent line meets the radius.

Point of Tangency (P.T.): The end of the curve, where the tangent line meets the radius.

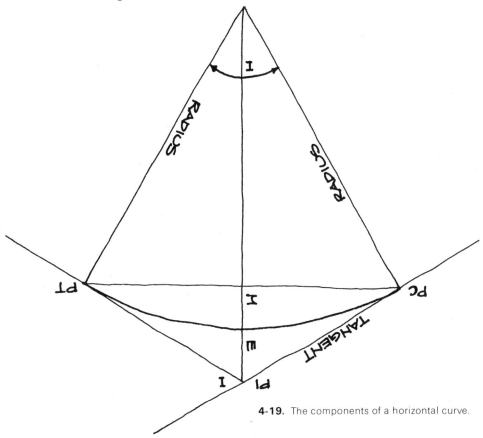

4-19. The components of a horizontal curve.

Point of Intersection (P.I.): The point where the two tangents intersect.

Included Angle (I): The interior angle of the curve, formed by the intersection of the tangents.

Long Chord (L.C.): A straight line between P.C. and P.T.

Middle Ordinate (M.O.): The distance between the center of the curve and the center of the long chord. The middle ordinate length plus the external distance is equal to the radius of the curve.

Deflection Angle (I): The angle at the change of direction between the two tangents. Equal to the included angle.

Degree of Curve (D): The angle at the center of a circular arc subtended by a chord of 100 feet.

Length of Curve (L): The distance from the point of curvature to the point of tangency measured along the curve for 100-foot chord lengths.

Radius of Curve (R): The distance from the midpoint of the circle of which the curve is a part and any point on the curve.

Several different equations are used to solve for circular curves. Some of these are:

$$D = \frac{5,730}{R}$$

$$T = R \tan \frac{I}{2}$$

$$L = \frac{I \times R}{57.29} \qquad L = \frac{I \times 100}{D}$$

$$I = \frac{57.29 \times L}{R}$$

$$L.C. = 2R \sin \frac{I}{2}$$

$$M.O. = R - \cos \frac{I}{2}$$

The calculations for horizontal alignment of curves are based on reducing the centerline of the proposed circulation system to a series of intersecting straight tangents, which are then connected with radii either according to prescribed standards or by trial and error. Often length-of-arc tables, figured for a radius of one foot, are used to compute *L*.

Vertical Alignment

Vertical curves are parabolic rather then circular, allowing vehicles to change gradually from an ascending to a descending grade (or vice versa) without bottoming out or getting hung up on the point of the curve. Calculations for vertical curves are also based on a curve joining two tangents. Following are the terms needed for calculating vertical curves (figure 4-20).

Point of Vertical Curvature (P.V.C.): The point where the tangent joins the curve in the direction in which the measurements are being taken.

Point of Vertical Tangency (P.V.T.): The end point of the curve, where the curve joins the tangent.

Point of Vertical Intersection (P.V.I.): The point of intersection of the two tangent lines.

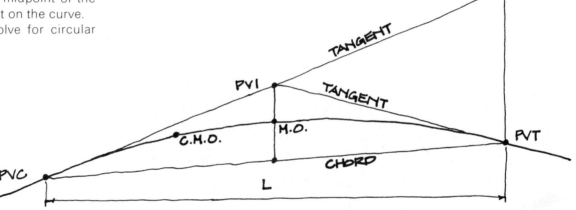

4-20. The components of a vertical curve.

Length of Curve (L): The length of the curved section joining the tangents.

Chord (C): A line drawn between the P.V.C. and the P.V.T.

Middle Ordinate (M.O.): The elevation of a point halfway between the chord of the middle ordinate and the P.V.I.

Chord Middle Ordinate (C.M.O.): The elevation of a point halfway between the P.V.C. and the P.V.I. measured at the M.O.

4-21. Details such as curb-cut construction and stairs determine accessibility.

4-22. A well-designed flight of perrons.

High Point (H.P.) or *Low Point (L.P.)*: The high or low point of the curve, obtained by multiplying the percent of the entering slope by the length of curve (L), and dividing this by the mathematical difference between the intersecting slope lines.

The quantity *d*: The algebraic difference between the intersecting slope (tangent) lines.

The variable *x*: Any horizontal distance from P.V.C.

The variable *y*: The vertical distance from the beginning tangent to the curve line.

In the calculation of vertical curves, downhill slopes are given a minus sign and uphill slopes a plus sign. Vertical-curve calculations are based on algebraic differences between points of elevation and percentages of slope. The only general equation used in the calculation of vertical curves is the one used to obtain the high or low point of a curve (a determination that is often necessary for drainage or sight distance reasons):[7]

$$y = \frac{x^2 d}{2L}$$

Accessibility

The detailing of the circulation systems adds to the users' perceptions of safety and quality. Foremost among the detailing considerations of designers should be accessibility: designing for access by the aged and infirm, and by the visually, audially, or physically handicapped; for wheeled vehicles, including but not limited to wheelchairs; and for anyone laboring under a temporary handicapping condition—whether being weighed down with packages, pushing a stroller, or experiencing back trouble. The design should invite everyone to use the site.

Detailing determines accessibility. Although providing a ramp or curb cut at each slight change in grade would appear to be a simple matter, such "access" points are often designed and located in such manner that all the trash and debris that washes down the street accumulates in them, denying the very access they were intended to allow. Another problem is that, while a gradual change in elevation allows users to change grade without having to negotiate a vertical step, it also creates a hazard for the visually handicapped unless a change in materials signifies that the sidewalk is ending and the street beginning (figure 4-21).

The controlled use of textured surfaces adds interest to any site and brings a welcome change from gray, brush-finished concrete. However, the way in which the surfacing is installed can again limit access, causing tumbles or twisted ankles as a person missteps or catches a heel.

When steps are included in the design, they should never occur as a single riser, but either in long perrons or as at least two risers in a flight. The proper height of the riser is related to the width of the tread. Exterior stairs are less steep and provide more foot room than corresponding interior stairs; therefore, they require a longer run for the same amount of rise. A standard formula for calculating the dimensions of exterior stairs is $2R + T = 26$, or two times the height of the riser plus the width of the tread is equal to 26. A riser under three inches high is dangerous because it is difficult to see; risers over six inches high appear and feel out-of-scale with the horizontal dimensions of the landscape. If a perron flight is designed with the object of serving as an unobtrusive change of grade, the color and/or texture of the nose of the tread should be changed to make the steps apparent to the viewer (figure 4-22).

A broad flight of stairs invites slow, unhurried movement and, depending on the orientation, is a favorite place of users to sit and enjoy nature. The narrower and steeper the flight is, the more difficult its ascent will seem to be. Landings should serve as resting places in extended flights of stairs, and should be designed to continue the sequence of experience visually rather than interrupting it.

The more places to sit there are, the better-used a site will be. The detailing of the stairs and ramps that provide easy access to a site can also provide stopping places for people—especially if the cheek walls are designed to allow sitting, or if the landings are large and casual enough to encourage spontaneous conversations.

The nodal development of circulation spaces minimizes the hard edge between circulation and activity space, allowing unpremeditated flow back and forth. If the seating and circulating spaces are well-lighted, enclosed but offering free and easy access to the main flow of traffic, protected from the sun and wind, and embellished with trees, shrubs, or other moving elements that add life to the site, they will be well-used and well-liked.

Storage

The detailing of a circulation system requires provision for the storage of vehicles when people arrive at their destinations. The general requirements of storage areas, whether for bicycles or for cars, are that they be within reasonable walking distance of the destination, light enough and open enough to allow visual scrutiny for safety reasons, and designed with adequate maneuvering space.

Reasonable Walking Distance

Reasonable walking distance depends on the amount of time people have to reach their destinations, on people's age and physical condition, and on weather. For most of the people most of the time, the closer to the entrance the parking space or drop-off point is, the more satisfied they are. A destination receiving a lot of quick-trip traffic will not be well-used if every two-minute errand requires a five- or ten-minute walk from a remote parking place. Most people can and will walk a block, however, if they have no choice.

The use of a site by the aged and infirm can be encouraged by designing convenient, close drop-off points for people who cannot or will not drive, and gradual slopes (preferably under cover) for those who can. Weather conditions decrease the distance people will walk between their vehicle and the entrance. The colder the wind is, and the harder the rain or snow, the more vehicles will be found as close to the entrance as possible, legally or not. Often a covered ramp is ideal from a weather-access standpoint but not from an economic or aesthetic one.

Bicycle storage is more difficult to control than the storage of other vehicles. Since bicycles are so maneuverable, unless the walking distance between the racks and the destination is very short and obvious, the rider will continue on to the closest tree or lightpost. As with cars, the closer the destination is to the storage area during inclement weather, the better.

Allowing Visual Scrutiny

Storage areas that are very enclosed and dimly lit will be perceived as unsafe by the potential users, in terms of the security of both their vehicles (from vandalism) and their persons (as they walk from the vehicle to the entrance). In addition, surveillance by security officers is more difficult in such areas, increasing the risk of vandalism or injury.

Adequate Space

Spaciousness and economy are closely (and inversely) related, with more generous storage areas usually costing more in terms of land assignment and surfacing required. Attempting to provide storage that falls below the minimum maneuverability limit, however, is false economy because a storage lot that forces the driver to work too hard to park will be used only when nothing else is available. Many designers fail to understand the need for adequate parking space for bicycles as well, assuming instead that all riders will be satisfied with the standard steel racks in which bikes touch and hit one another getting in and out. The investment that many bicycles represent is significant, and the owners are justifiably angry if their needs for secure storage are not recognized.

Parking Lots

There are many possible designs for parking lots, depending on turnover time, size of vehicle, and type of maintenance. The three different methods of parking—parallel, angle, or perpendicular—all have suitable applications.

Parallel Parking

Parallel stalls are the most difficult to park in—to the chagrin and frustration of people seeking their drivers' licenses. They also create safety problems, since they usually occur along the driving lane of a street or parking area. It is particularly difficult for a driver to park properly in a parallel stall on the left-hand side of the road, which happens on one-way streets. Even if the driver manages to get the car close enough to the curb, the wide-swinging doors of many makes of car can cause drivers in the adjacent lane to swerve abruptly—or even to hit the disembarking passenger or the car door as they drive by.

Angle Parking

The efficiency of angle parking depends on the degree of angle used. The closer this approaches 90°, the more efficient the lot becomes in terms of number of cars parked to amount of surfacing needed. Angle stalls are most easily entered and exited if the backing lane is designed wide enough to accommodate the occasional vehicle that does not pull all the way into the stall.

Both efficiency and ease of access drop if the lot is designed for two-way access to all driving aisles rather than for one-way access. Large angle lots should provide a two-way lane at either end of the rows of stalls, so that drivers searching for a parking place do not have to thread back and forth through every backing lane. The small triangular spaces left at the ends of the parking lanes provide good places for plantings, snow storage, or pedestrian safety landings.

Perpendicular Parking

Perpendicular or 90° parking is the most efficient in terms of number of cars parked to surface needed. However, in the interest of even more economy, designers will sometimes try to narrow the backing lane, forcing a "Y" approach into the parking stall and making it difficult for two cars to pass one another. Perpendicular lots are easy to lay out and easy to maintain.

General Design Issues

Any type of parking lot becomes dangerous without adequate sight distance at the ends of the driving lanes (to see oncoming cars) and at the intersection of the lot and the street (to see traffic without pulling into the street). Plantings designed to soften the harshness of large lots should either be high enough for drivers to be able to see under them, or low enough for drivers to be able to see over them. A square lot—rather than a long and linear lot composed of two rows of parking spaces separated by a single driving lane—is pleasing to the eye.

From the standpoint of convenience, it is desirable to arrange the lot with the driving lanes perpendicular to the front of the building, so pedestrians can traverse the lot without having to slip between parked cars. People will seek the shortest distance from their cars to the front entrance, and attempts to channel the pedestrian flow onto sidewalks should be made with the knowledge that people will use a walk that is parallel rather than perpendicular to their destination only if forced to do so by plantings or other barriers. Walks placed along the edges of parking lots and against the curb should be at least six feet wide to accommodate the overhangs of the cars as well as pedestrians. Figure 4-23 shows standard dimensions for various parking arrangements.

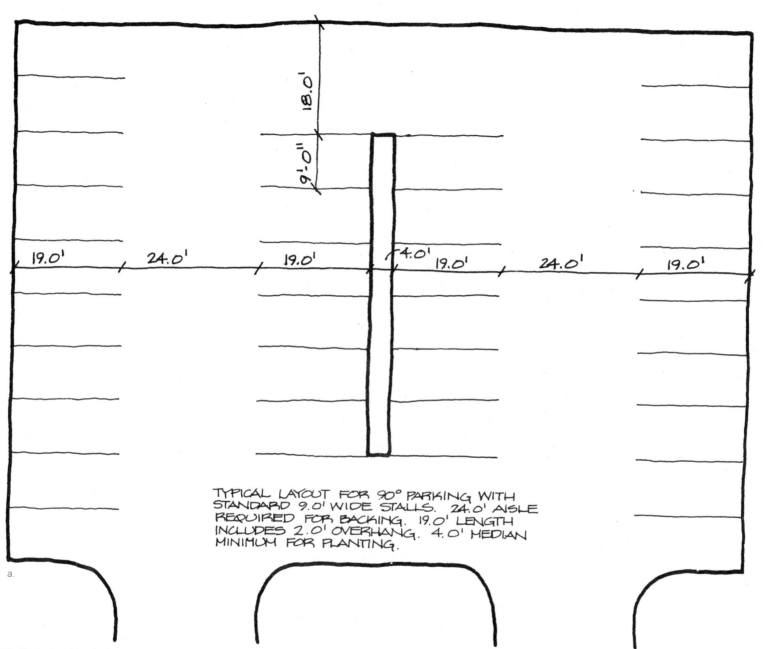

18.0'

9'-0"

19.0' 24.0' 19.0' 4.0' 19.0' 24.0' 19.0'

TYPICAL LAYOUT FOR 90° PARKING WITH
STANDARD 9.0' WIDE STALLS. 24.0' AISLE
REQUIRED FOR BACKING. 19.0' LENGTH
INCLUDES 2.0' OVERHANG. 4.0' MEDIAN
MINIMUM FOR PLANTING.

a.

4-23. Typical parking lot layouts: a. perpendicular parking; b. 45°-angle parking, one way; c. 60°-angle parking, one way; d. 45°-angle parking, showing
change typical for any-angle parking to allow two-way access to parking bays.

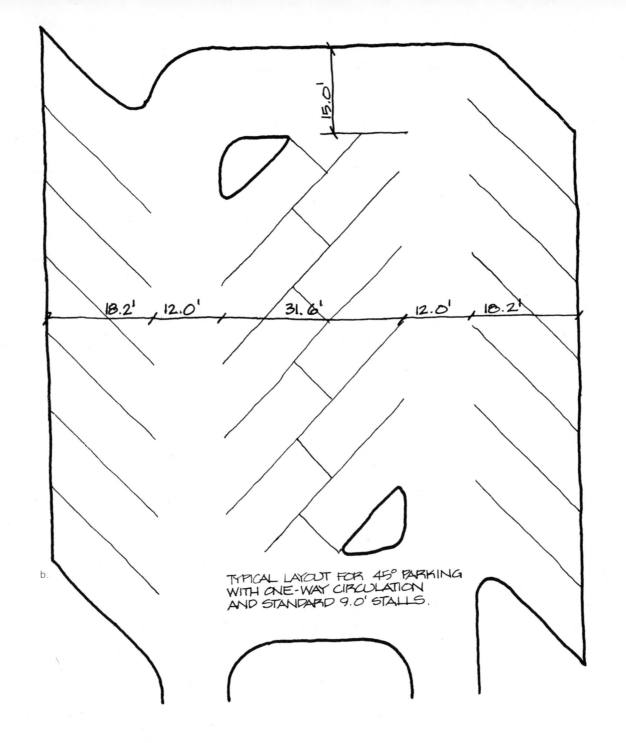

18.2' 12.0' 31.6' 12.0' 18.2'

15.0'

b.

TYPICAL LAYOUT FOR 45° PARKING
WITH ONE-WAY CIRCULATION
AND STANDARD 9.0' STALLS.

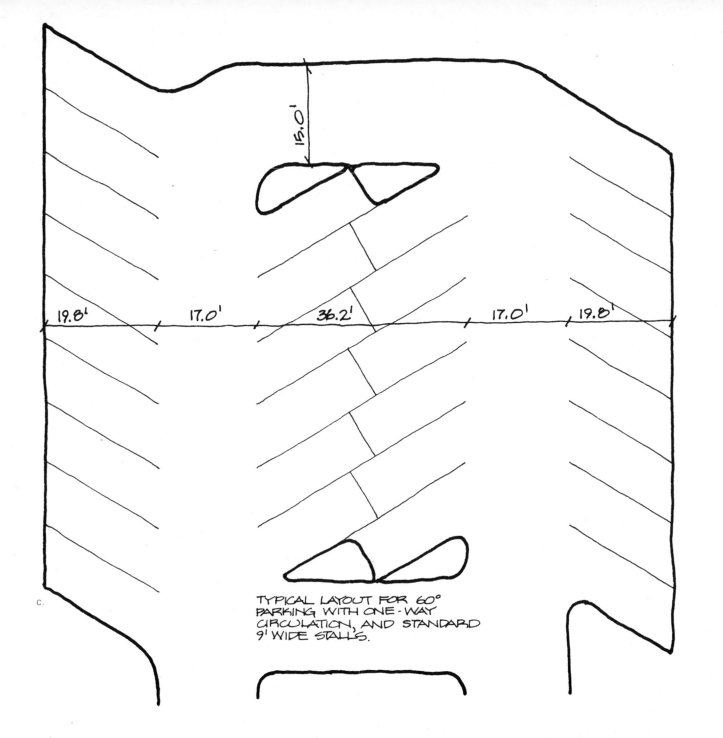

15.0'

19.8'   17.0'   36.2'   17.0'   19.8'

TYPICAL LAYOUT FOR 60°
PARKING WITH ONE-WAY
CIRCULATION, AND STANDARD
9' WIDE STALLS.

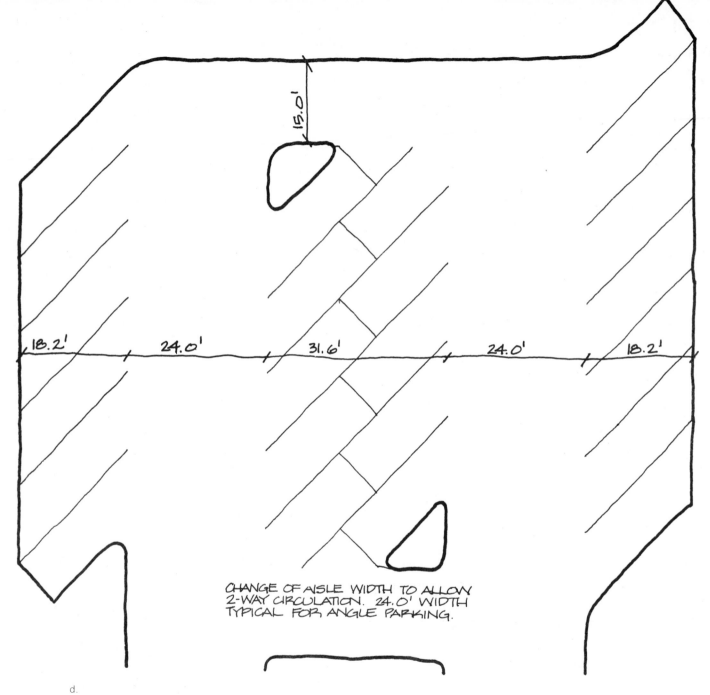

15.0'

18.2' 24.0' 31.6' 24.0' 18.2'

CHANGE OF AISLE WIDTH TO ALLOW
2-WAY CIRCULATION. 24.0' WIDTH
TYPICAL FOR ANGLE PARKING.

d.

REFERENCES

1. Kevin Lynch, *Site Planning*, 2nd ed. (Cambridge, Massachusetts: M.I.T. Press, 1971).
2. John Ormsbee Simonds, *Landscape Architecture* (New York: McGraw-Hill, 1961).
3. Simonds, *Landscape Architecture*.
4. Lynch, *Site Planning*, pp. 121–124.
5. Lynch, *Site Planning*, pp. 146–149.
6. American Society of Landscape Architects, *Landscape Architect's Handbook of Professional Practice* (McLean, Virginia: ASLA, 1972).
7. Russell C. Brinker, *Elementary Surveying*, 5th ed. (Scranton, Pennsylvania: International Textbook Company, 1969).

5 | **Plant Materials**

The subject of plant materials arises repeatedly in the design process: as a consideration during the initial steps of the process, as a prime determinant of enclosure in exterior space, as an indicator of climate and a solver of climate problems, and as a detail element in circulation systems. This chapter addresses the specifics of plant materials in the design of site, space, and structure.

DEFINITION OF SPACE BY PLANTS

Plant materials create a bond between people and nature, softening the sometimes thoughtless effects of progress and making even large cities more pleasant to live in. Plants may define space by creating masses or voids, by framing spaces or views, by acting as a backdrop to an interesting site feature, or by serving as a focal point. Plants define spaces psychologically as well as visually and physically.

Psychological Definition

The psychological definition of space provided by plants is somewhat difficult to describe, since people interpret spaces according to their own backgrounds, experiences, moods, and desires. Nearly any mood that the designer wishes to portray can be created by the careful selection and placement of plant materials. By combining different plant textures, colors, heights, and densities, a designer can make a space appear to be constantly changing, with places to suit any mood. A space featuring equally spaced plants of the same type can appear uniform to the point of monotony; the plants may then become a background, focusing attention on some other, more important aspect of the space.

The use of dark colors and fine textures that form a single plane when viewed from a distance can create a somber mood, as can using pines or spruces through which the wind whistles in the winter. In contrast, light-barked, delicate-leafed plants that rustle in the slightest breeze create a feeling of motion and flight. The introduction of color into the landscape often occurs through the use of flowering plants. The effect of coolness and quiet can be obtained by using cool colors in restful masses; alternatively the site can be splashed with a lively patchwork of oranges, reds, and yellows that encourages activity (figure 5-1).

Visual and Physical Definition

Both the psychological and the visual definitions of space are reinforced by the physical form the space assumes. However, spaces may be visually defined or suggested with very little actual physical enclosure. An example is the use of a change in the texture of a groundcover—from grass to something taller and coarser—to keep people off one area and on another. Placing a grassy area next to a concrete surface that is not separated by a change in elevation might suggest that the concrete is to be walked on but the grass is not.

As the visual definition of spaces becomes more obvious, so does the reinforcement provided by the physical qualities and placement of the plant materials. Plant materials occupy space in the same three planes that delineate and define space in general: the ground, canopy, and wall planes. As the height of the plant increases, so does its impact on the character of the surrounding space. The same is true of its density: the denser the material is, the more dominant it becomes. The character of a space defined by plants that are of uniform density at all levels is different from that of a space defined by plants with variable density.

The Ground Plane

Use of a solid, homogeneous mass in the ground plane or at a height below four feet creates a plane that either invites use (as when the material is a lush carpet of green grass) or discourages it (as when a dense, thorny shrub is planted). Use of a material of slightly different texture in the ground plane can reinforce the spatial activity overhead. For example, when a groundcover is planted under the canopy of a large shade tree, the ground plane's support of the canopy creates a bond between the two.

The Overhead Canopy

The overhead canopy of large shade trees, evergreens as they mature, and smaller, often-flowering understory trees creates spatial definition by introducing a plane (the underside of the canopy) at a level that is more comprehensible than the heights of surrounding tall buildings or even the upper height of the plants themselves. Many trees can be limbed up to create an umbrellalike feeling, with the outer reaches of the branches slightly descending and enveloping the people underneath. The shape of the plant determines whether the overhead canopy will be experienced as something under which one can sit or as a vertical barrier.

The density of the canopy, as with architectural overhead planes, determines the quality of the spaces underneath by affecting the shadow patterns. The lighter and lacier the leaves and branches, the less obvious the relationship between canopy and ground planes. A dark, solid canopy creates a solid shadow on the ground plane that in itself serves to define a space. Movement through a space can be suggested by placing the plant materials to delineate by their shadows the edges of the paths, rather than using a solid barrier of shrubs (figure 5-2).

The canopy becomes a scale-giver in large, impersonal spaces, such as downtown skyscraper plazas: the foliage of the trees stops the eye and limits the feeling that the buildings extend forever upward. When used in large open spaces, the canopies of the trees create enclosure by defining spaces under and between canopies; they also reduce the apparent area of the space.

a.

b.

5-1. Softening an edge: a. unobtrusive but contrasting background; b. reducing maintenance and adding uniform textural interest.

The Vertical Plane

Hedges and shrub masses are commonly used vertical-space definers. Large overstory trees spaced in uniform rows create an edge definition between areas because of the verticality of their trunks. They can be used in this way to reinforce a curve appearing in the base plane, or to outline and enhance a straight axis toward a terminal feature (figure 5-3).

Randomly spaced trees of varying forms can be unified in a single space that is defined by the trunks. The more uniform the textures and sizes of the trunks, the more they will read as the repetition of a single element. A different approach is to choose plants with very interesting bark or trunk structures, in which case the vertical lines of the plant become more interesting than either the canopy or the ground plane.

The overall form of the plant (as well as the trunk) may be vertical, drawing the eye upward. A tall, narrow, columnar plant reads as a unified linear element, whereas a broad-spreading overstory tree reads as a vertical trunk element and a horizontal overhead canopy.

Plants with vertical forms or with a height that exceeds their spread can be used in the landscape as reference points or landmarks. A single tree or grouping can be a focal point in a large space. A group of symmetrical, vertical evergreens can be used to contrast sharply with a horizontal form or an open landscape.

Plants can be used in three general ways in spatial design: as individual specimens that are allowed to assume their full size and natural habit; as monocultural masses, in which the uniform use of a plant material creates a solid form that acts as a backdrop or boundary; and as mixed masses, in which plants of different forms, colors, textures, and degrees of seasonal change attract attention, direct it elsewhere, or complement other elements or spaces (figure 5-4).

5-2. Shadows define space.

5-3. Vertical trunks define space.

5-4. Three ways of using plants: a. specimen; b. monocultural mass; c. mixed mass.

a.

PLANTS IN CLIMATE CONTROL

Plants play a major role in the design of exterior space through their ability to control climate. The first clue to the climate of a site occurs through observation of the plants growing there, since plants not only control but also indicate existing climatic conditions. Since most plants have a fairly specific set of growing conditions that must be met if they are to develop into healthy specimens, the presence or absence (and health, if present) of certain plants tells the designer something about the texture, density, fertility, and composition of the soil, and about the amount of light present, the temperature extremes, and the moisture distribution at the site.

The plants present on the site may be native or introduced. Native plants are indigenous to the site area, naturally reproducing themselves without benefit of human interference. Introduced plants are native to some other part of the world but have been successfully established on the site. While the native vegetation is perhaps more indicative of general climatic conditions, the introduced plants may actually be more sensitive to minor

b.

c.

variations and thus may give a more detailed picture of the microclimate.

Plants that withstand a certain set of conditions on one site should withstand the same climate elsewhere. The plants of the windswept plains, for example, should manage to survive in other hostile environments with similar temperature variations, wind patterns, soils, and moisture.

Wind Control

Plants are extensively used to control wind. Any plant that is not completely horizontal or so open in structure that it is barely visible will control the wind to some extent. Efficient wind control is usually obtained through a massive planting of evergreen and deciduous materials at a place between whatever is to be protected and the prevailing winds; however, even a single plant offers some protection, as becomes clear in observing livestock huddled near a tree or a person standing in the shelter of an evergreen while waiting for a bus.

Heat Control

The heat gain and loss that occur on a site are controlled by using plants to produce shade, to reduce or channel wind, and to insulate structures. Shade on the exposed faces of a structure or on the surface of a playground slide will reduce the amount of radiant heat gain, thus keeping the effective temperature in the shaded areas cooler than the temperature in full sun.

In the winter months, plants that provided cooling shade in the summer meet the need for increased radiation by obligingly dropping their leaves. When planted against a foundation, plants create an air space that is not affected by climatic conditions as severely as open space is and that does not allow heat or cold to escape from the structure as readily. Dark green leaves reduce glare and heat reflected into surrounding spaces and structures, as well as having the psychological effect of soothing the eyes, especially in the urban environment.

Precipitation Control

Anyone who has sought shelter from the dripping rain under a tree canopy (*not* under a single tree, which is a great lightning attractor) knows the shielding effect of the interlaced twigs and leaves. Some plants are so dense that they allow moisture only from the longest soaking rains to reach the ground immediately below.

Precipitation in the form of snow or ice is controlled by changing drift patterns by means of windbreaks and snow fences. When placed properly, these living snow fences can reduce much of the manual labor associated with shoveling snow and chipping ice.

PLANTS AS TRANSITION ELEMENTS

Plants are a primary material in the development of transition spaces between the hard, manufactured materials of structures and pavements and the natural environment. They are used to soften the lines of buildings as they enter the ground plane and to reduce the apparent expanse of large paved areas. They can screen poor views or direct attention elsewhere. A plant can be treated as an architectural element, repeating the form of a manufactured one. Plants added to a location characterized by too much architectural diversity, can, if carefully selected, give continuity of color, texture, or form to the space. They temper the city's abutment with the rural landscape and the wilderness by lending softness and life to all three environments.

Reinforcement of Design Character

Plant materials can be used to create transitions between spaces by reinforcing the design character of the structural elements. The choice of shape, color, size, and texture of the plants should be made with the appearance of the structures in mind. Gray or buff concrete can be contrasted with a purple-leafed plant; the dark color tends to tone down the glare of the concrete, and yet the contrast is greater than it would be if a green plant were used (often a pink tone in concrete receives emphasis from purple leaves).

Because of the huge variety of plant materials available, a plant can be chosen for every odd corner that needs character, or plants can be used en masse to screen an undesirable view. By enclosing the spaces behind and adjacent to a structure, plant materials can create a pleasing background for prominent architectural features. The profile of a low building can be given a sense of height by using vertical plant forms to lead the eye upward. A tall building can be given human scale at the entrances by the use of horizontal plant forms.

a.

b.

c.

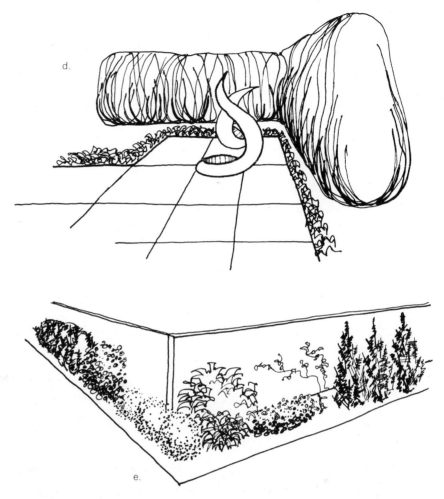

d.

e.

5-5. Plants can be used in many ways: a. to create a background; b. to give height to a low structure; c. to create human scale; d. to give a short building length; e. to create an interesting view.

Plants can also be used in the ground and wall planes, set at a slight distance, to create lines leading to the entrances of the structure. A short, blocky architectural form can be given length by trailing plant materials from slightly in front of the structure to either side of it. A blank wall can be used as a canvas, with various plant materials creating patterns of texture, color, and shadow against it. An overhead canopy can be used to create a sense of shelter and invitation (figure 5-5).

The View from Above

Often the reinforcement of structural elements on a site is considered only in terms of the exterior appearance of the building, not from the viewpoint of those living and working within. The latter consideration may actually be more important, however, because a person visiting the site on an occasional basis will experience the spaces afresh, while an office window framing a single portion of the site may offer variety only according to the plants and daily weather conditions. A miniature landscape can be created especially for viewing from a window—with variations in scale, texture, and color that might even play perspective tricks on the site and lengthen or shorten the horizon—or the structure can be made to focus on an interesting site view or vista, thereby creating a visual relationship between the structure and the site.

5-6. Topiary spirals, and an elephant in the background.

Treating Plants as Architectural Elements

The treatment of living plants in an architectural manner to match or blend with manufactured materials is still another way to create a comfortable transition between site, space, and structure. The designer must recognize, however, that plants treated in this way must be hand-selected for certain design characteristics; in addition, the maintenance capability and budget must be adequate to ensure that the plants remain in their artificial forms.

Hedges

The most common architectural treatment of plant materials consists of clipping them into square or triangular hedges of uniform height and spread. Such hedges are used to define the edges of paved or planted areas, functioning in much the same way as fences or walls composed of manufactured materials. Such hedges, when clipped to heights above eye level, can be used to separate two spaces from one another totally. When used at or near building foundations, hedges may extend the apparent line of the building outward, or they may direct attention to an entrance.

Topiary

Another architectural treatment of plant materials is a type of pruning and training known as topiary. This is an exaggeration of the clipped hedge, in which plants are trained into whimsical animal or geometric shapes—elephants, for example, or spiraling cones. Evergreen materials work best for this type of treatment. Topiary should be used as one would use site sculpture, since the plant will be seen not for its natural characteristics, but for its created shape (figure 5-6).

Espalier

Two other types of pruning—espalier and pollarding—are also used to create special effects. Espalier combines severe pruning with rigid training of the young stems of the plant so that it assumes a flat, two-dimensional character. A classic candelabra or modified pyramid is often used as a model for espalier. The plant continues to flower and fruit as though it were three-dimensional, but it does so in a flat plane. Espaliered plants are especially effective when seen against a bare wall, where the geometry of the training can be appreciated. Like topiary, espalier involves extensive maintenance (figure 5-7).

Pollarding

Pollarding is the continual pruning of a plant at the same point on the branch year after year, producing a large, stubby callus and giving the plant a coarse, gnarled appearance. When in leaf, a pollarded plant often looks as though it has been topped. While a single specimen may be pollarded, this technique is most effective when used on a uniform large planting to create an orderly, manufactured appearance.

Pleaching

Pleaching is the interlacing of the branches and twigs of trees, either naturally or artificially, to form a tight overhead canopy. Many overstory trees pleach themselves naturally (especially when closely spaced) and form graceful arches over city streets and pedestrian walkways. Artificial pleaching is more common in Europe—where the "pleached allée" constitutes a strong design feature of many sites—than in the United States. The cool shadiness of the space beneath pleached trees truly seems like an outdoor room (figure 5-8).

Maintaining Architecturally Treated Plants

While the architectural treatment of living plant materials can be an effective way of creating a transition between built and unbuilt environments, the designer must be fully cognizant of the design characteristics of the plants being used before specifying them as a design feature. Expecting a fast-growing hedge to be clipped as needed, when that may be as often as once a week during the main growth period, is sometimes unrealistic. In addition, plants chosen as matched pairs (for use in symmetric designs or to create overall uniformity in a hedge or other mass planting) will not necessarily remain matched as they grow. Therefore, a design including architecturally treated plant materials must be able to accommodate minor inconsistencies without losing its integrity.

THE TERMINOLOGY OF PLANTS

An understanding of the basic terminology of plant materials is essential before the designer can begin to work with them in any detail. Following is a list of common terms and their definitions.

5-7. Espaliers against a wall.

5-8. A naturally pleached "allée."

Overstory: The canopy or highest type of cover. Produced by the larger trees (at maturity, thirty feet tall and more). Most often, overstory trees are used for shade in energy conservation situations; they can also help bring the scale of large vertical elements down to the human dimension. Most are deciduous, although very old evergreens can assume the characteristics of overstory trees.

Understory: The intermediate level of trees or large shrubs, found naturally under the canopy trees. Height range in the understory is from twelve to thirty feet. Many common ornamentals and blooming trees are found in this category; they can be used for focal interest, or as a gradual means of bringing the scale of larger trees down, or as a visual screen or backdrop.

Evergreen: A plant that retains its foliage year-round, but distinguishable from a "broadleaf evergreen" by its needlelike or scalelike foliage. Height range is extremely variable, from prostrate to over two hundred feet. Form is also variable, although many people think immediately of the pyramidal Christmas tree as the form of evergreens. Evergreens are very valuable in climate control, in providing living interest during the winter months, as a backdrop for ornamental plantings, and as a screen or traffic director.

Conifer: A cone-bearing plant, in most cases synonymous with "evergreen."

Broadleaf Evergreen: A broadleaf plant of variable height and form (many are in the mid-sized shrub range) that retains its green, leafy foliage throughout the year. In the cool and temperate macroclimatic zones, the foliage of broadleaf evergreens, though retained, does not remain predominantly green, but turns shades of bronze, brown, purple, and gold. The leaves of broadleaf evergreens are thicker than those of deciduous plants and frequently are covered with a waxy coating to help prevent desiccation during cold, dry, windy winters. Many groundcovers are broadleaf evergreens. They are useful for providing winter interest and year-round screening, and they provide a good textural contrast to the needled evergreens.

Shrub: A plant distinguishable from trees by its generally smaller size (height is from twelve inches to about twenty feet) and by its growth habits. Shrubs are multistemmed from the base or ground, whereas trees have a single stem unless intentionally injured or otherwise forced to put up more than one stem.

Shrubs are variable in form—from weeping to round to columnar, and everything between. They are excellent for establishing human scale, for providing seasonal interest, for screening undesirable features, and for creating enclosure.

Groundcover: Any plant that covers the ground. Types of groundcover are chosen for their capacity to spread rapidly and tightly, for their relative ease of maintenance, for their ability to exclude weeds, and generally for their habit of low growth. The term groundcover may include small shrubs (most designers do not consider a shrub over twenty-four inches in height a groundcover) that sucker profusely, vines, bulbs, perennials, broadleaf evergreens, or annuals. Groundcovers other than grass are excellent for use in places where grass is difficult to maintain, such as very steep slopes, rocky sites, odd corners that would otherwise demand trim mowing, planters, and places where a subtle means of controlling foot traffic is desirable.

Annual: A plant that sprouts, grows, blooms, sets seed, and dies in a single year. Plants that are perennial in one zone may often be successfully grown as annuals in another. Annuals are used extensively for bright splashes of color; they are available in heights ranging from a few inches to over ten feet. Forms vary from upright to rounded and they may spread or vine. Because of the need for replacement every year, annuals are considered to make somewhat high maintenance demands.

Biennial: A plant that puts out vegetative growth the first year, flowers the second year, and then dies and must be replaced. Biennials are used most often in perennial gardens where the vegetative growth of the first year can be hidden among other blooming plants.

Perennial: A plant that persists from year to year, usually dying back to the ground or sending up new shoots from the rootstock each spring; it is distinguished from a shrub by the lack of woody growth. Perennials can require extremely high maintenance (as with hybrid tea roses) or almost no maintenance, as with native perennials. In selecting perennials, the designer accepts lower maintenance, but also a blooming season that is generally shorter than that of annuals. Perennials may provide color from early March through October.

Vine: A plant with a distinct trailing or climbing habit and usually characterized by very fast growth. Vines may be annual, perennial, herbaceous, or woody. Some shrubs will assume vine-like characteristics when grown near a vertical surface. Vines supply quick cover and can provide shade in a relatively narrow space.

Deciduous: Losing leaves on a seasonal basis. Deciduous plants are highly useful for solar control, but they need the added density of evergreens if used in a windbreak. If the plant has interesting bark, form, twig patterns, or fruit, it may continue to provide interest after its leaves have fallen.

Herbaceous: Containing little or no woody tissue (leading most herbaceous plants to die to the ground in the colder climates). Most perennials are herbaceous; very few shrubs are.

Hardiness: A plant's ability to withstand the climatic conditions of the area in which it is located. Hardiness determines the amount of maintenance a plant requires, the potential success of the planting project, and its economic cost. A complete determination of hardiness is based on success not only in handling broad general patterns of wind, sun, and temperature, but also in handling pollution, soil conditions, the presence of unusual concentrations of chemicals, excessive acidity or alkalinity, competition from other plants, and so on.

Native: A plant that is indigenous to the region in which it is found. The native plants were present on a site before it was disturbed; they tend to be the hardiest plants.

Introduced: A plant not indigenous to a region, but native to some other area of the world. Introduced plants have proved adaptable to conditions other than those of their native sites.

Naturalized: An introduced plant that has succeeded on its own to the extent that it has reseeded itself and become "native." Plants that have escaped cultivation or that have been allowed to spread at random are described as naturalized.

Exotic: A somewhat controversial term used to describe plants not native to a region, but also suggesting questionable hardiness for the site involved. "Exotic" seems to connote the lush foliage of the tropics and of hothouse plants, although this is not necessarily true.

Genus: A level of differentiation in the scientific identification of living things. Plants of the same genus have similar basic characteristics, and each member of the genus will have the same first name in the two-part scientific name of each plant. All oaks belong to the genus *Quercus*, and all maples belong to the genus *Acer*.

Species: A more specialized level of scientific identification: the species name is the second name in the scientific name. The species designation differentiates one oak from another, for example, or one maple from another. The northern red oak is *Quercus borealis*, while the white oak is *Quercus alba*; and the Norway maple is *Acer platanoides*, while the amur maple is *Acer ginnala*.

Variety: A further categorization, even more specialized than species. Identifications of varieties are usually based on refinements of a given characteristic, such as (in plants) flower color or growth habit. The "Pond" Norway maple *(Acer platanoides "Pond")* differs from the common Norway maple in the thickness and color of its leaf.

PLANT CHARACTERISTICS

As is the case with all design, form follows function in the use of plant materials. Each species of plant has its own unique design form, and each cultivar or variety of that species has a unique form as well. The characteristics that differentiate one species from another make a plant perfect for one use and unsuitable for another. These distinguishing characteristics are reliable indicators of the species: an oak bears acorns; an apple tree, round red, yellow, or green fruit; and a pine tree, cones of some sort.

The characteristics that may be common to members of a given species include: overall form; mature size and shape; bloom color, appearance, and timing; bud size, shape, and placement on the stem; similar leaf attachment, venation, form, and edge; typical branching patterns; discernible bark characteristics (such as roughness, peeling, or color); growth habit (such as upright, spreading, or suckering); and a tendency to turn a certain fall color. All members of a species retain the same overall characteristics: a Norway maple will remain a Norway maple regardless of its variety. The varietal difference may be in the form, color, longevity, or size of the plant's blooms, or it may be in the leaf size or color. Consider, as an example, there are several hundred named varieties of flowering crabapple.

As a species, the crabapples are small, spreading ornamental trees with oval green leaves and fragrant, five-petaled flowers that bud in shades of pink, opening to white, followed by fruits called *pomes*. The variety called "Royalty" has an overall shape that is uniform and rounded, with purple foliage and deep magenta blossoms, followed by almost black fruit. "Red Jade" is very dwarf, has profuse, pendulous stems, and is covered with persistent red fruits well into the winter. "Spring Snow" overall is large and round, its blooms are white, and it does not fruit. Yet all of these—different though they are—can properly be called flowering crabapples.

Individual Variability

Even though there are characteristics common to a species, individual specimens are infinitely variable. Because of individual variability, a row of pines, all planted at the same time and originally the same size and shape, comes to assume the picturesque form shown in figure 5-9. The differences that appear among plants of the same species (or variety, for that matter) may be the result of differences in the location, in the parent plants from which the present plants were grown, in the climate and growing conditions, and in the general treatment received.

5-9. Individual variation.

Northern-grown plants exhibit higher cold tolerance and slower rates of spring bloom than do their southern-grown counterparts, which often have a lush appearance and softer foliage and twigs.

A sugar maple in one microclimate may reach a height and spread of eighty feet, with a full, round crown and a brilliant yellow fall color; another individual of the same species in a different location may have an open crown, reach a height of only forty feet, and turn orange in the fall. Willows that tower in the air and weep gracefully at the edge of a pond in the temperate macroclimatic zone become stunted, tangled groundcover in the frozen tundra of the far north.

For design purposes, an individual plant of a given species can be expected to exhibit design characteristics similar to those of other plants of the same species. Unless uniformity between individuals is critical to the design, it should not be necessary to hand-pick each plant. If a particularly important relationship between the plant and the spatial and structural qualities is being sought, however, hand-selection may be needed.

INFLUENCES ON DESIGN QUALITIES

The design qualities of plants are affected by the plants' surroundings, by the quantity and quality of light reaching the site, by the numbers and placement of the plants—whether they are apparent as individuals or are seen in clumps or masses—by the distance from which the viewer experiences them, and by climatic changes.

Surroundings

There are four key aspects of surroundings that affect the choice of plant materials: scale, texture, complexity, and color.

Scale

The scale of a plant in relationship to its surroundings is what determines whether the plant will seem in proportion with the spaces and structures around it or out of proportion—either larger, overwhelming them, or smaller and being overwhelmed (figure 5-10).

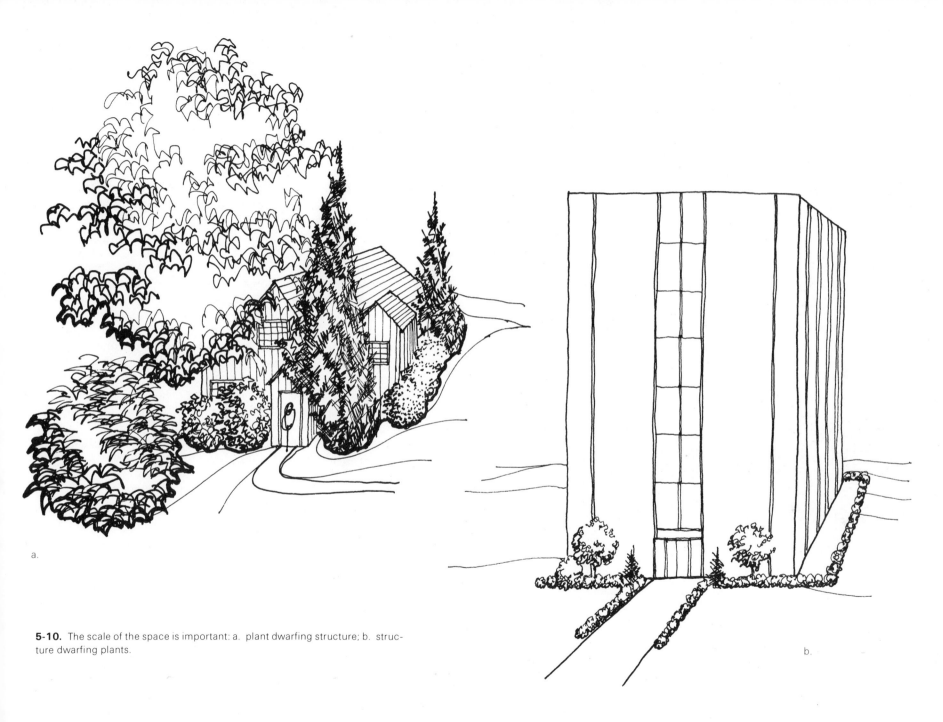

5-10. The scale of the space is important: a. plant dwarfing structure; b. structure dwarfing plants.

Texture

Since texture affects our perception of the scale of an object, the texture of nearby buildings can exaggerate a proportion problem or correct it. A very coarse-appearing structure, with unevenly spaced openings of varied size, will not be enhanced by being surrounded by a large variety of coarse plant materials that possess interesting design characteristics of their own. A smooth barren wall, on the other hand, may need all the living excitement it can get to attract attention to the building.

Complexity

A building with complex nooks and crannies and different elevations, roof lines, angles, and corners may need to rely on plants of uniform shape placed at critical points to unify the visual form of the building. Each site and structural situation is different, demanding individual treatment.

Color

The colors of nearby buildings in relation to the colors of the proposed plants—and in relation to how the latter colors change throughout the year—is often overlooked or considered in terms of only one season. This results in hot pink blooms against bright orange brick, or yellow flowers or fall colors against concrete that has an obvious purple tinge. In such a case, the whole visual quality of the space clashes.

In their zeal to provide interest at all seasons of the year, designers often try to create a sequence of bloom, from early spring to late fall. Unfortunately, the colors that are to follow one another in sequence are not always carefully chosen, and since the exact blooming time for each plant and species varies with climatic conditions and site peculiarities, the fading, washed-out pinks may be seen at the same time as the blazing yellows or vibrant blues.

As with the use of too many textures, shapes, or sizes, too many colors or too much of one color creates confusion rather than focusing visual interest. The infinite shades of green available by combining various plant materials can be used to tie a planting plan and site plan together, with an occasional splash of contrasting color added for interest. Since most plants are in bloom or in fall color for a relatively short period of time, the choice of green can be more important than the choice of flower or autumnal color.

Light

The amount and intensity of light in which a plant is seen affect the shadows it creates and can accentuate or reduce its apparent depth and density. The leaves of many plants are so fine and thin that the slightest ray of light shining behind them produces a soft, translucent green that almost seems shadowless. The thicker and darker the leaf is the less light will penetrate and the greater the contrast between light and dark areas will be on the site. Cloudy days tend to wash the color out of bright plants, reducing the contrasts and creating a more unified appearance. Very bright days also wash the color away, producing a hazy, gray-green appearance.

Spacing

The number of plants of a particular species, their size, and their spacing in relation to other plants and to open spaces change the appearance of a site considerably. If a designer uses five plant varieties, all differing in one or more significant traits (such as form or growth habit), and spaces them equally, the user will see each plant as an individual and will not create a balance between the plants. Two or more plants of similar characteristics placed near each other will be read as a group, but will still be visible as individuals.

The more plants of the same type are used in a design, the more homogeneous the design becomes (figure 5-11). Careful juxtaposition of forms, textures, colors, and sizes against one another can create focal points within a plant mass, or can direct attention to the center or the ends, or can lead the eye upward or out. Two elements equally spaced produce a tension that is difficult for the viewer to resolve (figure 5-12).

Distance

The distance at which a plant or plant mass is seen affects its character and quality. The farther away a person is, the more uniform the appearance of the plants becomes, and the finer the texture seems. As one moves closer, the individual plants in a mass gradually become identifiable, and then the details of each individual plant do, too.

Climate

Just as plants influence and indicate the microclimate of a site, so does the climate affect the design characteristics and general appearance of the plants—on a daily or even hourly basis as well as seasonally. Certain plants release a stonger scent during humid, hot weather, which changes a person's perception of the site. Other plants may close their flowers during either the heat of the day or as evening approaches. Still others rustle in the slightest breeze, adding sound to the site experience.

In the cool and temperate macroclimatic zones, plants go through a distinct period of seasonal dormancy, with deciduous plants dropping their leaves and evergreens ceasing growth. A plant chosen specifically for multiseasonal interest might have a colorful or sweet-scented bloom in spring, followed by lustrous, shiny foliage during the summer, a rich, red autumn color, and a unique branching habit or bark texture during the winter.

Many plants possess above-average interest during at least two seasons. A plant can change from fine to coarse texture with the seasons, or from a dark, dense mass to open branching.

5-11. Spacing determines visual quality: a. equally spaced plants of different forms read as specimens; b. similar forms begin to relate; c. form variety, yet continuity.

5-12. Two ways of arranging focal elements: a. a central focal element; b. competing elements.

What is dominant during one season may be subordinate or even absent during another: the bright twigs of the redtwig dogwood are barely visible when the plant is in full summer foliage, but they stand out sharply against snow or evergreens during the winter.

LIVING VERSUS MANUFACTURED

Plants differ in a number of ways from manufactured or hard materials. An understanding of the differences is necessary to ensure that each material is used properly, to its best advantage.

Change

One significant difference is that plants are dynamic materials, constantly changing with changes in level of daylight, with the seasons, and with age and maturity. The "Christmas tree" planted at the corner of the house and decorated with lights for ten or fifteen years gradually loses its neat symmetrical shape and small size, becoming a fifty-foot giant without any lower branches. The seedling that appears too spindly and weak ever to produce shade for the south side of a building becomes substantial enough to hold several small children in fifteen years. The eventual mortality of plants, too, is rarely considered during design, leading to landscapes that mature and decline without benefit of any young, strong replacement plants to fill the gaps.

Patience in Visualizing the Design

The constant changes that plant materials undergo are hardly ever experienced by the designer, who may never see the site again after finishing the project. Design with plant materials therefore demands patience and the ability to visualize a site in ten, twenty, or fifty years. Architectural or manufactured construction is something that can be finished. Designer and client alike can feel confident that, once the structures and pavements are constructed according to the plans and specifications, the architectural job is complete. The landscape is never finished, however, and neither is the site that the landscape defines: both continue to change over the years.

Individuality

Because of their machined nature, architectural materials tend to be so similar to one another that a person would have trouble telling individual bricks or wood members apart. Not so with plants; each plant is an individual. This leads to a certain unpredictability that must be considered when designing the site. Two shrubs picked for their shape and size and placed symmetrically as a gateway to the site may bloom two weeks apart one year and on exactly the same day the next. Since plants depend on the cues of the actual weather to begin their seasonal transformations, budding may start early or late in the calendar spring, and leaves may hang on well into November one year, only to be coated with ice and dropped on the first of October the next. Two seedlings from the same tree may grow to be completely different-looking plants. The only constant of which the site designer can be sure is change—day-to-day, season-to-season, and year-to-year.

SPECIFYING DESIGN CHARACTERISTICS

The choice of plant materials for a site can be made without knowledge of the scientific names, if the designer understands how to choose characteristics to fulfill the needs of the site and realizes that written specifications will be needed to control the variety, size, quality, method of planting, and guarantee. It is far better to describe plants according to these characteristics than to attempt to specify plants in a site plan using only the English names. As an example, consider the common name, "red maple." When people ask at a nursery for a red maple, they usually want one with dark red summer foliage. Several maples have this foliage variation, however, and these are not really "red maples" but varieties of Norway maple.

Determining General Character

In planning the use of plant materials for a site, the designer must first determine the desired overall effect in schematics. Is the site to appear uniform, with the plants acting as screens and backdrops for the activities, spaces, and structures in front of them? Is the site to be composed of a number of small, separate spaces, given continuity by the use of plant specimens that are similar yet unique? Is the site to be a tapestry of textures, colors, forms, and sizes, to be examined at leisure, or is it to be observed at high speed from a great distance? Are the primary users of the site people who could easily be injured by careless attention to detail (for example, children who might run afoul of thorned shrubs in their play, or elderly people who might be endangered by trees that drop slippery fruit on walkways)? Are the plants' sound, smell, and touch as important as their visual qualities to the site's overall character?

Dominance

Once the general character to be sought is ascertained, the designer determines which elements should dominate in order to produce that character—the space, the structure, the landscape, or the views or vistas. The location and character of the dominant elements, the character and scale of the dominant spaces, and the details needed to exert dominance are decided at this point. Plant masses and specimens are located, and their general form is outlined. Elevations or perspective sketches might be drawn to match the three-dimensional structures and spaces with the plants. Finally, the design characteristics of each plant are chosen.

Detailed Design Characteristics

The detailed selection of plant materials begins with selection of the plant *forms* that will best fulfill the needs of the site—for masses, specimens, or a combination of both.

There is a plant form suitable for every site need. The form of the plant determines whether it will stand out as a specimen or blend with its surroundings. A soft, rounded form contributes to a gentle, rolling site character, whereas distinctly upright or pyramidal forms grab and hold the viewer's attention (and must therefore be used sparingly). Forms can flow up the side of a building, or they can crawl and sprawl over the edge of a wall. Following is a description of commonly found plant forms. These and others are shown in figure 5-13.

Vaselike: Ascending lower branches, turning in a cascadelike form toward the top of the plant. Some of the stateliest shade trees and a large proportion of shrubs are vase-shaped.

Pyramidal: Cone-shaped, with a central leader. The branching may be horizontal, weeping, or ascending. Pyramidal plants are formal and symmetrical; they are useful as an accent plant or for screening.

Prostrate: Hugging the ground, as flat as possible. Groundcover, useful as a mat or transition from other shrub plantings to grass, is the primary prostrate form.

Fastigiate: Exhibiting an extremely narrow, upright habit, with a central leader. Fastigiate forms are rarely wider than six feet; they serve as accent plants—very formal, useful in narrow plantings or as a screen or sculptural element.

Weeping: Branching is decidedly descending and fountain-like, with or without a central leader. Weeping plants are useful as specimen plants and for transitions between walls and greenspaces or between taller plants and the edge of the ground plane.

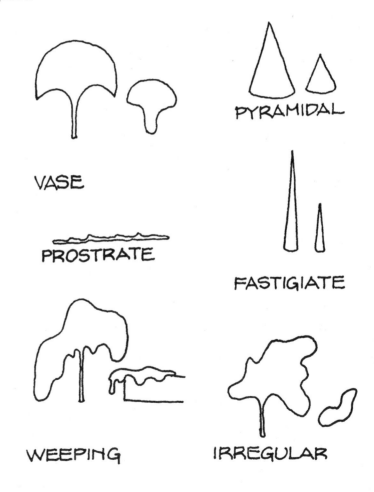

5-13. Common plant forms.

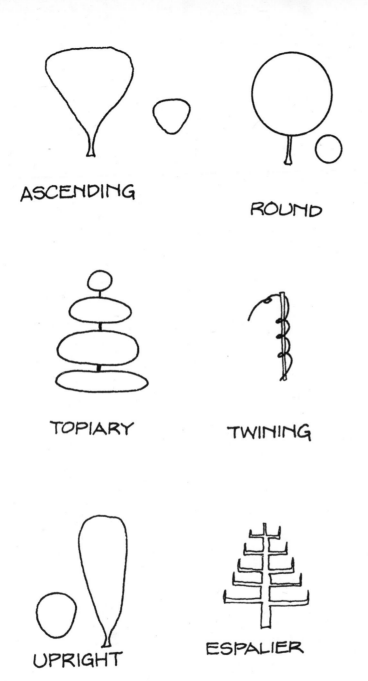

ASCENDING

ROUND

TOPIARY

TWINING

UPRIGHT

ESPALIER

Irregular: Open crown in overstory; rather amorphous shape in shrubs. The irregular form may combine many attributes of other forms. Because they are informal and casual, irregular plants are useful in masses and naturalized settings.

Ascending: Branching is angled upwards, giving the plant an almost top-heavy character. A tree with ascending form provides shade and can be used as a street tree whose canopy extends over traffic and pedestrians.

Round: Globular (as wide as or wider than it is tall) and often very dense. Round plants may be formal or informal, depending on pruning treatment, and may be used to bring the scale of narrower elements down to human dimension.

Upright: Almost completely vertical in character, although the plant need not necessarily be narrow. An upright form makes a good transition to rounder shapes and may be used to accentuate an opening or a vertical architectural element.

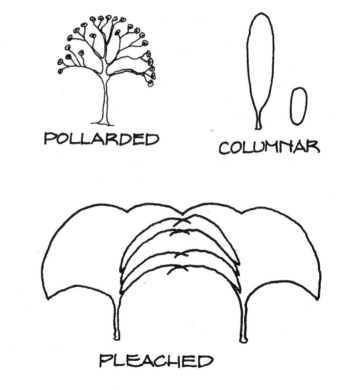

POLLARDED

COLUMNAR

PLEACHED

Columnar: Narrow, but not as narrow as the fastigiate form, with an upright overall appearance. Columnar plants are useful to supply accents, to line approaches, or to provide a formal backdrop or screen.

Twining: A growth habit in which the plant twists around an object or stem. Twining plants are useful for covering unsightly walls, fences, or posts.

Clinging: A vine growth habit in which the plant fastens itself to a surface by means of small suckers or aerial roots.

The overall plant form visible from a distance helps determine the general character of the site. The forms of the detailed elements of the plant—its leaves, flowers, fruit, and twigs—contribute not only to the overall form, but to specific, focused interest.

Leaves can be smaller than a fingernail or as large as a dinner plate; they can be round, elliptical, triangular, or heart-shaped, straight-edged, toothed, or lobed, blunt or pointed, thick or thin, waxy or dull, simple or compound, and dark, light, or variegated. The leaf imparts much of the textural quality to a plant during the growing season, so choosing it is important to the overall design (figure 5-14).

Although the flowers and fruit of plants are significant for only short times every year (in most cases), they add such visual interest that plants are often selected on the basis of their flowering or fruiting characteristics alone. The form of the flowers may be round or trumpet-shaped, single or double, and may occur in long strings, in panicles, or in tiny clusters. Fruits vary from the fine cottony seed clusters of poplars to the winged samaras of maples to the large globes of apples and pears to dry papery capsules.

The branching structure of a plant gives it its form, making the winter appearance of a deciduous plant similar to its appearance when clothed with leaves. The very ends of twigs or whole branches can droop or can come to a sharp point. Plants can branch and rebranch to form dense clusters of twigs. The twigs themselves can be smooth and fine, or foursided and angular.

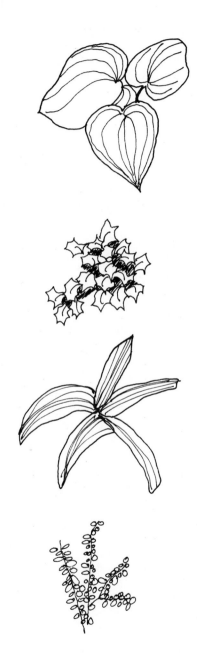

5-14. Textural differences in leaves.

Additional Design Characteristics

The desired color, hardiness, smell and sound, texture, seasonal change, and growth and maintenance requirements for each plant should be determined, in addition to size and form.

Color

The design characteristic of color should be given significant weight in the choice of plant materials. Color variation occurs from plant to plant, even in plants of the same parentage. A prime example of this can be seen in a field of Colorado blue spruce: out of any group of seedlings, only a chance few will have the marketable blue color sought by commercial nurseries.

Not only should the color of the leaves and flowers be considered, but that of the bark and twigs as well. The Scotch pine with its distinctly cinnamon-colored bark, the redtwig dogwood, and the white birch all might be chosen for their interesting bark color rather than for some other design quality.

Hardiness

Hardiness is a design characteristic that overrides all considerations of aesthetic quality when choosing plant materials. No matter how beautiful its blooms or form, if a plant cannot withstand the climatic conditions of the site, it is an unsuitable choice. Most references list plant hardiness by zone (figure 5-15). In specific situations, a plant may be hardy in one or more zones north or south of the optimum locations, with adequate protection from hot sun or drying winter winds.

Sheltering microclimates can be constructed on a site to meet the needs of a semihardy plant. The designer must recognize, however, that the more questionable the plant's ability to survive in a given situation is, the more maintenance will be required to keep it growing in satisfactory condition, and the more responsibility for proper care and cost will fall on the client. The choice of a less interesting but hardier plant is wise in most situations because inadequate maintenance of a potentially beautiful plant will result in an undesirable design.

Hardiness is measured by a plant's ability to withstand the macroclimatic and microclimatic extremes presented by a particular site. These include extremes of heat and cold, seasonal change, sun or shade, wet or dry soils, heavy clay, sand, or loam, high or low humidity, competition from other plants, reflectivity and pollution of urban environments, continued pruning and clipping, and snow and salt buildup. A plant that survives in a cool climate with a lot of snow cover but little humidity may not be able to handle climatic conditions in a warmer zone with high humidity.

The unpredictability of the weather from year to year adds another set of problems; a year of severe drought followed by a period of excessive rainfall will tax nearly any plant's capacity to survive. Many plants are so sensitive to requirements for certain soil, shade, or moisture conditions that they will thrive happily in one location on a site but weaken and die in another location just a few feet away.

Smell and Sound

Smell and sound are two design characteristics that are not often considered in choosing plant materials, but they add still another dimension to the experience of the site. Scent is not limited to blooming flowers; certain trees and shrubs emit odors from their bark, leaves, or fruit as well—and not all of them are pleasant, as is evident from such common names as "Skunkbush," "Carrion Flower," and "Stinking Ash." A sweet-smelling variety of plant may seem oppressive and cloying if it is present in large quantities, overpowering the other site experiences.

Sounds are produced in plants according to the branching structure and the manner of attachment of leaf to twig. The dense but fine-needled structure of evergreens allows the wind to whistle and wail through their branches, producing what to some people seems a melancholy sound. Cottonwoods and other poplars move in the slightest breeze, producing a "white noise" type of background sound. Some deciduous trees hold their (nolonger-green) leaves well into the winter months, producing still another kind of sound—a dry, hollow rustle.

Plants not only produce their own sounds to compete with other background noises, they attract wild animals and birds that bring additional song and chatter to the site.

Texture

In balancing the textures of the various plants used (and the relationships between living and manufactured textures), the designer weaves a tapestry. Plants of similar texture can be placed in proximity to one another to create an overall effect of mass. Contrasting textures, as long as their contrast is not so severe as to create dissonance, direct attention to and lead the eye to focus upon a particular place.

Texture is a comparative design element when used with plants, just as it is when used architecturally. In other words, a relatively fine-textured plant only seems fine-textured if it is seen against a blank background or a background of a coarser texture. Textures should be chosen for their dominance or subordination. Texture in the landscape is often meant to be physically experienced; the arrangement of plants of varied textures so that they can be touched will allow the site to be enjoyed by people lacking one or more other senses.

Seasonal Change

The primary concern in choosing plants for their appearance at different seasons relates to the degree of independent enclosure needed on the site. Although a shift in color is the seasonal change most widely noticed by people, the changes in texture and density are the ones that can inadvertently turn a private retreat into a public showplace if the designer is not careful.

Growth and Maintenance Requirements

Growth and maintenance requirements vary from plant to plant and depend also on the desired finished appearance of the site. The detailed growth characteristics of each plant determine

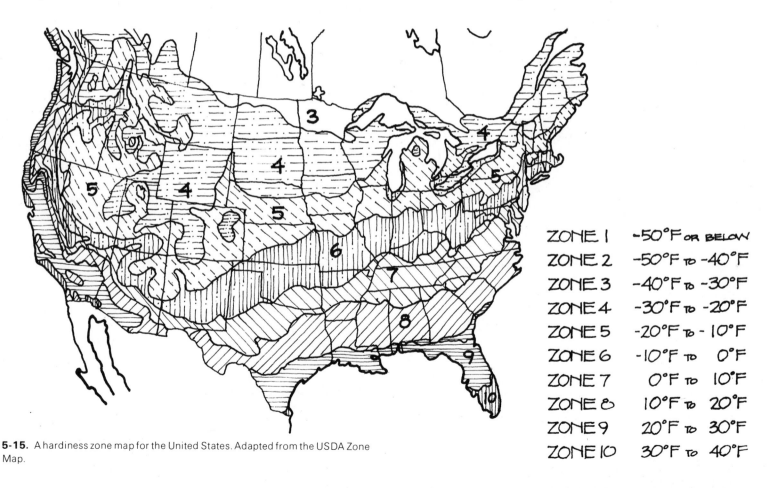

5-15. A hardiness zone map for the United States. Adapted from the USDA Zone Map.

ZONE 1   -50°F OR BELOW
ZONE 2   -50°F TO -40°F
ZONE 3   -40°F TO -30°F
ZONE 4   -30°F TO -20°F
ZONE 5   -20°F TO -10°F
ZONE 6   -10°F TO 0°F
ZONE 7   0°F TO 10°F
ZONE 8   10°F TO 20°F
ZONE 9   20°F TO 30°F
ZONE 10   30°F TO 40°F

how much maintenance will be required for it. The designer should consider the plant's root structure and habit, top growth, flowering, fruiting, and leaf characteristics, and all seasonal variations thereof.

In terms of root structure, plants are either fibrous-rooted or tap-rooted. Large-size fibrous-rooted plants are often easier to move than tap-rooted plants of equal size. However, many fibrous-rooted plants are surface feeders (with their roots very close to the surface of the ground) leading to competition with groundcovers and shrubs and resulting in bare ground or in heaving and buckling pavement. Some plants have root systems that seek out and penetrate any water or sewer line, making maintenance difficult and expensive (figure 5-16).

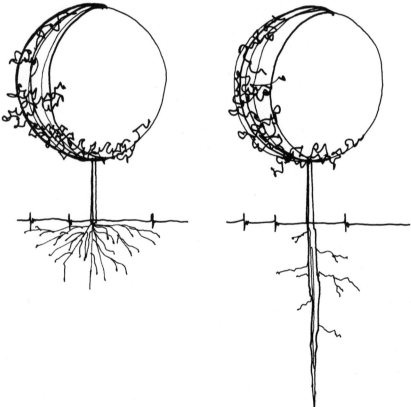

5-16. Fibrous- and tap-rooted plants.

The characteristics of the plant's top growth constitute a major maintenance consideration, since the form of the plant is one of the first design characteristics considered. The general growth habits of the plant determine whether it is suited to the space available. Use of a wide-spreading shrub next to a sidewalk raises the necessity of constant pruning to keep the branches from causing problems for pedestrians. An overhanging plant placed too close to a structure may cause damage to the roof as its branches brush against it. Leaves clogging gutters and downspouts can pose a maintenance problem. A narrow, ascending tree may have internal growth problems that require constant pruning to keep it healthy.

The particular characteristics of a plant also determine how easy it is to maintain. Maintenance people may be reluctant to prune a dense thorny shrub; and some groundcovers form such tangled mats that it is nearly impossible to remove rubbish and leaves from them—regardless of the tool used. In areas where there is a possibility of damage or injury from falling limbs, the designer should avoid the fast-growing trees, which tend to be weak-wooded and susceptible to ice and wind damage.

Seasonal litter is a consideration with almost all plants. The dropping of faded or wilted flowers sometimes creates a maintenance problem (especially when the flowers are large), but the fruit drop is usually worse. Flowering crabapple trees that drop their slippery fruit on a terrace or sidewalk, cottonwoods whose "cotton" clogs air vents and screens, and the many trees that litter the ground with quantities of spiky burrs are examples of plants that present fruit-drop problems. Another problem attends plants with prolific seed production, depositing and sprouting an annual crop of little seedlings in window wells, gutters, and sidewalk cracks. Some plants are not self-fruitful; that is, they are either male or female, or they need pollination by a different variety to produce fruit. If the fruit is deemed an undesirable characteristic, the male or nonfruiting plant should be chosen.

Availability and Cost

The availability of a plant in a particular region may constitute a design consideration, since cost and maintenance are affected if a plant must be brought from a great distance. A plant that is readily available locally can be replaced easily if the need arises.

This is not to say that special plants should not be sought for special effects, but the backbone of the design should consist of plants that can be readily purchased.

Cost is a factor in all design areas, including choice of plant materials. The uniqueness of the plant, its size, and the condition in which it is purchased all affect its price; additional costs hinge on the amount of labor required to install and maintain it through the guarantee period. Generally, the larger and more nearly perfect or true-to-form the plant is, the more it will cost.

Size also affects the way in which a plant can be moved. Plants may be purchased bare-root, balled and burlapped, or in a container (figure 5-17). Some cannot successfully be moved in bare-root form above certain sizes; others become root-bound so quickly in their containers that they become permanently stunted. Some plants cannot be moved once they reach a certain size because of deep taproots or delicate feeder systems. Others cannot tolerate fall planting: they need the summer months to reestablish root contact with the ground. The conditions of access to the site might disallow use of a large tree spade or introduction of the heavy weight of a balled and burlapped tree, necessitating the purchase of smaller plants. A designer who is unaware of special planting requirements imposed by particular plant characteristics or site conditions may find constant replacement and eventual substitution the unpleasant consequence.

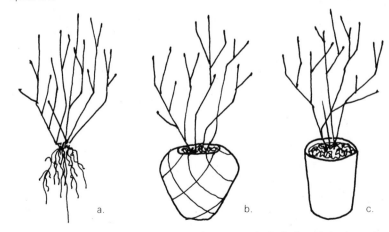

5-17. Ways to purchase a plant: a. bare-root; b. balled and burlapped; c. containerized.

It is the designer's responsibility when specifying plants to inform the client about how those plants will look. Most clients are unfamiliar with costs and sizes of trees, and they usually expect a large tree for a little money. Faced with a limited budget, the designer might suggest purchasing the major trees in larger sizes and forgoing some of the smaller materials until another year, since the trees are the framework of many planting designs and take longer to amount to anything than do shrubs.

ENERGY CONSERVATION DESIGN
Windbreaks

The most critical requirement of any plant being considered for use in a windbreak is hardiness—hardiness above and beyond that needed in practically any other situation (the exceptions being the harshest of urban environments). The hardiest plants are those that are indigenous to an area or that have proved themselves adaptable through a long trial period. Plants used in windbreaks must be able to withstand not only the common microclimatic conditions of soil and moisture, but the direct, full force of the winds they are intended to control. Many windbreaks, particularly in large-scale or rural developments, are given little special maintenance after their initial establishment. They must therefore be able to produce roots and top growth and survive times of drought or extreme cold, poor soils (without benefit of additional fertilizer), and too much or too little snow cover during the winter.

A plant's rate of growth must be viewed in connection with its hardiness—in particular, its ability to withstand storm winds without breaking. Since it is desirable to produce an effective windbreak as soon as possible after planting, plants with a medium-to-fast growth rate are more desirable than equally hardy but slower growing plants. Unfortunately, fast growth rates often mean weak wood.

The optimum design for a windbreak enlists plant materials of mixed species, with a variety of shapes, sizes, and densities. Overall, the plants should be dense enough to reduce the direct flow of the wind, yet open enough to allow partial penetration of the wind. Combining evergreen and deciduous materials creates this type of control, but because few hardy evergreen can tolerate shade, the use of overstory trees or even shrubs that will shade the base of the evergreens will eventually result in the thinning out and possible loss of the evergreens. Windbreak evergreens should be chosen for their mature shape, since with age many of them lose their conical, branched-to-the-ground structure and become in effect overstory trees. The increased wind speed that occurs when wind is channeled under such a windbreak can be avoided by combining shrubs with trees and large evergreens from the start.

There are two basic ways to lay out windbreaks: in a straight row or in a more random pattern (figure 5-18). Straight-row windbreaks consist of several rows of plants of different species, all of the plants in a given row being of the same species. The initial advantage of this type of windbreak is the ease with which plants can be established in rows, especially when there is no variation of species within each row. This makes the spacing simple to determine, allows water to be given to one type of plant but not to others, and provides for ease of cultivation between rows and plants (particularly if the spacing is such that mechanical equipment can get between the rows). Disadvantages of the straight-row windbreak include its rigidity of form (which makes it difficult to blend with a more natural environment) and the amount of space it requires. Another problem is that, if a particular species is stricken with disease, a whole row of the windbreak is endangered—although the row system also tends to make disease treatment somewhat easier.

The random windbreak mixes species within the entire area covered by the windbreak, allowing a more aesthetically pleasing overall appearance to develop, but perhaps sacrificing predictable control of wind and snow. The initial layout and planting of this type of windbreak is time-consuming, since the location of each plant—rather than just the ends of the rows—must be staked. Future maintenance can be difficult also, because the plants have to be located and treated individually. However, the random-pattern windbreak allows the designer to treat this energy-control feature as an integral part of the overall design, blending it into existing plantings and devising transitions between heavily planted and open spaces. Each species can be engaged for its design characteristics as well as for its ability to block the wind.

The spacing of individual plants in either type of windbreak depends on their mature sizes and on the density of the planting mass desired. However, in both cases the spacing of the plants should be triangulated; that is, instead of lining the plants up in both north-south and east-west directions, the spacing should be staggered to baffle the wind (figure 5-19).

A design that depends in part on a living windbreak for energy conservation should provide for future replacement and reconditioning of that windbreak when necessary. The windbreaks first planted in the United States no longer truly control the wind, having long since been limbed up, shaded out, and damaged by storms without being replaced. A design specifying removal and replacement of plants over each ten- or fifteen-year period will ensure that the windbreak remains viable.

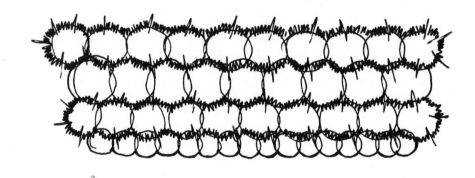

a.

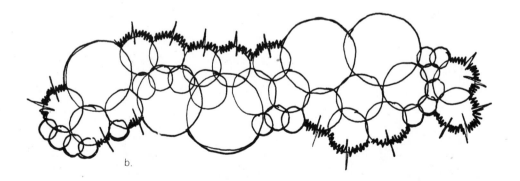

b.

5-18. Two windbreak designs; a. straight-row; b. random.

Snow Fences

Planning a living snow fence can be done either separately or in conjunction with wind control. Plants in snow fences must be just as hardy as windbreak plants, and must also be able to withstand being almost covered by snow. While breakage is not quite the same problem as it is with windbreak plants, longevity is: most shrubs are not long-lived. Shrubs used for snow fences are quite fast-growing, providing control in three years or less. A mix of species will provide varied density to direct the wind's penetration and (consequently) the location of the major drifting.

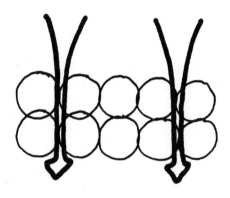

Shade, Moisture, and Temperature Control

The use of plants to provide shade and control of microclimatic conditions of moisture and temperature is widely recognized. The placement of individual plants or plant masses is very important in this type of control, and should be governed by specific site conditions and by the design characteristics of the plants being used. In the cool and temperate macroclimatic zones, deciduous plants are used for shade, because their wintertime dormancy allows the sun's rays to enter the site during the underheated months. The actual amount of shade provided by a plant depends on the plant's structure, density, and shape.

Although wide-spreading overstory trees are usually the first ones that come to mind for their shade capacity, plants exist that will provide shade suitable to virtually any design situation. Very narrow spaces, for example, may comfortably accommodate columnar trees or shrubs, or even vines. Since the branching structure of many trees becomes higher with age, a design for consistent shading might include stratified planting or smaller plants placed beneath the canopy of larger ones to fill gaps left by the trunks of the larger trees. This not only helps the overall shading of the site, but proves aesthetically more pleasing than when plants are placed equidistantly—with no overlap of canopies (figure 5-20).

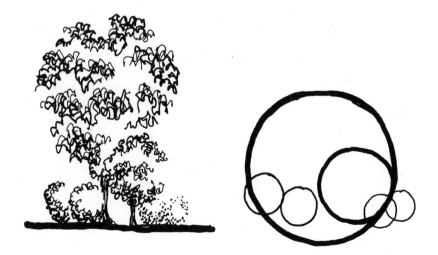

5-19. Triangulated spacing blocks the wind.

5-20. Stratified plantings produce more shade.

LOW-MAINTENANCE PLANTING DESIGN

The key to successfully designing a site for low maintenance lies in proper assessment of site conditions and proper choice of plants to meet those conditions. Even the hardiest native will not survive if planted in soils or given moisture or fertilizer to which it is not suited.

Analysis for Low Maintenance

The analysis of a site to reduce maintenance must take into consideration the amount and type of use each site area will receive. High-traffic pedestrian areas can be planted with grass if the client is willing to spend the extra time and money required to care for it properly; if the client is not prepared to do so, however, a hard surface is in order. Reducing high-maintenance areas to a minimum by surfacing or planting in masses allows labor-intensive types of work to be concentrated in a few small areas.

Although the same design characteristics (color, form, texture, balance, and scale) are considered for high- and low-maintenance designs, low-maintenance plans are usually simpler—relying on form, texture, and repetition rather than on great variety. If the shapes of land and planted areas are kept simple, with variation occurring in the leaves and stems of the plants, less maintenance will be required to keep nonplanted or mowed areas in shape. Repeated individual use of the same type of plant, or grouping plants in masses rather than using isolated specimens, translates into fewer specific maintenance requirements and more general maintenance.

Plants suitable for a low-maintenance design should be chosen for their hardiness and for their ability to stay in bounds—that is, to maintain a desired shape without much pruning, to refrain from spreading, or to keep from outgrowing the size of the space allotted to them. Unless plants are specifically chosen for their natural forms, and then allowed to grow, bloom, and produce seed naturally, the site will not be as low-maintenance as it could be. A plant poorly chosen for a given location and then continually pruned to keep it in bounds is likely to lose all of its natural beauty, including the formation of its flowers and seeds.

Generally, the greater the proportion of the site that is allowed to become naturalized (whether through succession and the gradual spreading of plants on their own or in a designed way that is calculated to look natural), the less work that will be required to maintain it.

Plant color and interest are achieved not only through the use of annual flowers and of plants that require yearly attention to live through seasonal changes, but also through the selection of woody plants and masses of groundcover. The low-maintenance or natural landscape is subtler than an architecturally treated landscape; it is more casual and freer in its design, suggesting rather than directing motion or visual focus. The more native plants are used, the greater grows the likelihood of successful naturalization.

In the quest for low-cost, low-maintenance landscapes, many sites are being converted from their previous high-maintenance designs. Often, plants can simply be allowed to revert to their natural state; that is, pruning and other maintenance can be stopped. In other cases, additional planting (to reduce the amount of grassed area to be maintained) or removal of high-maintenance plants may be necessary. Lawn areas may be converted to native grasses, which, once established, thrive on the natural climatic conditions of the region with little additional attention. A site that is to be established in natives from the start will take longer than one using introduced materials, because a characteristic of the hardiness of many native plants is that they put down their supporting, nutrient-supplying root structures one year, before showing much top growth. The site may therefore look unkept or even unplanted for the first two years or so.

Many times the conversion of a high-maintenance site to low-maintenance can be accomplished simply by explaining to the client or caretakers what the actual requirements of the site's vegetation are. Pruning, fertilizing, watering, and other maintenance chores can all be done to excess when based on incomplete knowledge of the plants' needs.

ARCHITECTURAL TREATMENT OF PLANTS

On the other end of the maintenance spectrum is the architectural or formal treatment of a site.

Plants chosen for architectural treatment must appear totally uniform when used in a mass, as a hedge, or as a background, or else must be chosen for their qualities as individual specimens. Architectural treatment of a site rarely includes naturalization or playing the characteristics of one plant against another. Plants

5-21. An architectural planting.

that spread rapidly, that will not stay neatly in bounds, or that have unpredictable growth habits subject to change with maturity are not well-suited to this type of design.

A client requesting landscaping with a manicured, geometric appearance should understand that a high degree of maintenance is usually required or alternatively that a large proportion of the site will have to be surfaced with low-maintenance hard materials. As a comparison of the labor required to take care of certain types of plants, bluegrass requires ½ hour per 100 square feet per year, annual flowers require close to 3 hours per 100 square feet per year, and roses require almost 17½ hours per 100 square feet per year.[1] Although comparable figures for native turf are not available from the same study, a conservative estimate is under 5 minutes per 100 square feet per year. In any case, there *is* a considerable difference in maintenance and cost between a highly manicured, totally maintained landscape and a natural one (figure 5-21).

REFERENCES

1. John VanDam, John W. Mamer, and William W. Wood, *Labor Inputs by Type of Plant per Year* (University of California, Division of Agricultural Sciences).

6 | Grading and Drainage

Grading and drainage constitute a critical functional and aesthetic part of integrating site, space, and structure. Unless a building is to be set directly on an existing foundation, at least enough grading will be needed to ensure drainage from the building outward. Thus, nearly every site will require some grading or drainage work.

FUNCTIONS, SPECIFIC USES, and AESTHETICS

The primary reason for grading any site is to provide drainage paths that will keep unwanted water away from elements of the site that might be damaged or made unusable or hazardous by it. Besides this purely functional reason, however, a site may be graded to allow for some specific use or to enhance the site's aesthetic quality.

Function

Grading to make a site functional can be classed according to one of five purposes: to allow for drainage, to provide access, to make more of the site usable, to reduce maintenance needs, or to preserve existing features.

Drainage

Drainage is an inherent consideration in all other functional reasons for grading a site. It is obviously critical around buildings intended for human habitation or use (because lack of attention to the land surrounding a building can create flooding problems that endanger life and ruin property), but drainage must also be provided to avoid such climate-related problems as icing or ponding in parking lots that are too flat, and wind- or water-erosion of susceptible, overgraded soils. Unless proper grading for drainage is used, plant materials needed for climate control and landscape enhancement will suffer from saturated or excessively dry soil conditions. A playing field, playground, or other recreational development will not be functional if its grading either allows too much water to run across the site or provides so little drainage that ponding results (figure 6-1).

Access

Adequate access for pedestrians and vehicles is a primary functional requirement of most sites. Pedestrians and vehicles can negotiate flat or slightly sloping grades easily, but steep grades with difficulty or not at all. The users' physical characteristics and abilities—their age, condition, familiarity with the site, and degree of mobility—and the mechanical abilities of the cars, trucks, vans, motorcycles, or bicycles using the site determine the degree of slope that can be specified in the grading plan to maintain a functional site. Ease of access directly affects the amount of use a site will receive. Awkward or confusing access routes will discourage use. In fact, this sort of design, which might be called "detraction" instead of "attraction," can be used purposely to keep people and vehicles out of certain areas. An example is the trails systems into the fragile areas of some national parks. Because access to these areas is made difficult

and steep, fewer people enter them than would if the access routes were obvious and easy.

Making More of the Site Usable

Terracing and contour farming, common practices in many farming regions of the world, are examples of grading to make more of the land usable—in this instance by reducing the severity of the slope. This type of grading can also be used in urban situations. Terracing (changing an area of constant steep slope into a series of shorter extreme slopes separated by large flat areas) can provide space needed for structures, open space, parking lots, or recreation fields. However, the extensive regrading needed to construct large terraces can be very costly, in terms both of dollars spent and of irreversible impact on the land and natural site features (figure 6-2).

6-1. Grading to provide drainage also reduces erosion. Courtesy USDA Soil Conservation Service.

6-2. Parallel terraces to make more of a site usable. Courtesy USDA Soil Conservation Service.

6-3. Without retaining walls, steep slopes suffer serious erosion and require frequent, expensive maintenance.

6-4. Regrading will be needed to preserve this creek. Courtesy USDA Soil Conservation Service.

Reducing Maintenance Needs

Grading to reduce the amount of maintenance required by a site is a logical consequence of grading for function, access, and the reduction of slope. Low maintenance needs bear a direct relationship to adequate drainage. Sites that drain properly will not contain abrupt transitions that cause ponding, scalping, or similar maintenance problems. Regrading can lessen erosion problems on steep slopes; reduce the number of odd corners that cannot be reached except with hand equipment; and limit adverse impacts of climate (figure 6-3).

Preservation

A final important functional reason to grade a site is to preserve valuable existing features. These may include shrub masses and trees, rock outcroppings or other geological forms, streams, ponds, or other water features, and man-made elements such as roads, parking lots, or sidewalks. Preservation grading for plant materials involves not only maintaining the original grade elevation under the dripline of the plant, but also ensuring that the same amount of water reaches the plant after grading is finished. When left to their own devices, streams change location, carving deeper and deeper into the earth and meandering from side to side, and ponds silt up or seep away into the surrounding soil. Grading to preserve these water features also contributes to lower maintenance requirements and a more aesthetically pleasing site (figure 6-4).

Commonly, a site is redesigned because of a change in use, and new features must be designed to share the site with useful existing features. Preservation grading is necessary in such projects to maintain the elevations of existing features that are to remain—whether they be parking lots, terraces, roads, or sidewalks.

Specific Uses

Highly structured or defined outdoor activities often depend on the fulfillment of specific grading requirements if they are to function properly. The designer should be aware of these specific activities from the start of the grading process because (as with the circulation system) grading for these functions may determine the remainder of the site layout. The development of playing fields, courts for tennis, basketball, and other sports, ski runs, resort areas, running, cycling, and riding tracks, and out-door amphitheaters and seating terraces are just a few special uses that may require exact grading. Not only must the site accommodate the specific grading requirements needed to serve these functions, but the transitions to less rigid areas of the site must make visual sense and be smooth and functional.

Aesthetics

Finally, but no less importantly, grading is done for aesthetic enhancement of the site and its features. The sensitive, skilled designer can go far beyond merely satisfying the functional and specific use requirements of most sites by varying the final appearance of the land through grade manipulation. The fine grading of a site contributes as much to the overall site quality as do the relationships of balance and proportion between open and enclosed spaces. Different reactions to a site are generated depending on whether the approach is downhill or up, whether the focal points are seen against a backdrop of hills or across a flat plane, whether the playground rises in a series of levels from the land surrounding it or stands in stark contrast to the environment.

Grading can turn a flat landscape into an exciting experience of hills and valleys, thereby creating anticipation about what lies around the next bend, or it can convince a user that a path is not as steep or as long as it really is. The eye can be made to follow the land directly up to the walls of a building through grading, extending the apparent height of the architecture, or a site can be graded to appear as a gently descending slope when in reality it is composed of a series of terraces that are visible from the downhill side only (the English "ha-ha" concept).

Plant materials needed to screen an undesirable view can be given an instant height boost by being placed on a man-made berm three or four feet tall. The land can be shaped to orient the view outward, offering the viewer the most distant horizons and instilling in the viewer a sense of the infinite (and also a sense of humility); or the land can be bowl-shaped and introspective, psychologically if not physically shutting out the world. Grading plays a major part in the development of all characteristics of a site and, consequently, of the human attitudes toward it.

ANALYSIS FOR GRADING

Whatever the reason and ultimate goal for grading a site, the process begins with a thorough understanding of the program requirements and with an analysis of the existing grades and site features affecting the final design.

During site analysis, both the obvious, visible surface features and the less obvious subsurface features should be examined. In addition, the designer should look closely at the conditions of the adjacent properties.

Surface Features

Analysis of surface features relates mainly to the degree and direction or orientation of slope. Examining the degree of slope will indicate to the designer which areas can be developed with the least financial or physical impact. Slope orientation determines to a great extent the climatic conditions of the site; as the degree of slope increases, favorable or unfavorable site conditions ensue because of orientation.

Degree of Slope

Natural and man-made slopes range from completely flat to nearly vertical. Slopes of less than 2% are considered flat for surface-graded grass areas. Slopes of 33% are difficult to mow, and most slopes over 45% require stabilization. A list of minimum and maximum slope guidelines is given below:

1% Absolute minimum for concrete and grassed areas; ponding will usually result; avoid if possible.

1.5% Minimum for asphalt; workable minimum for concrete.

2% Workable minimum for grassed areas and asphalt without ponding; gradient visible against a fixed horizon; minimum away from buildings for first 15–20 feet.

3% Gradient at which slope becomes obvious.

5% Easily negotiable slopes for cars, either up or down; preferred maximum for first-class city street work.

6% Too steep for most semitrailer trucks without changing gears and slowing down.

7% Maximum grade for parking lots in both directions; too steep in climates that receive a lot of ice and snow.

8.33% Maximum grade for access ramps for the handicapped.

10% Sidewalks noticeably difficult to climb and dangerous in ice conditions. Acceptable for some road work if absolutely necessary.

25% Optimum maximum for mowable slopes.

33% Absolute maximum for mowable slopes.

50% Maximum for slopes held by a groundcover; and depends on soil stability.

In landscape work, slopes are commonly given either as percentage figures representing the change in elevation for each one hundred linear feet, or as a ratio of run to rise. For example, a slope of 10 feet of rise per 100 feet of run would be shown as a 10% slope, or as a ratio of 10:1 (figure 6-5).

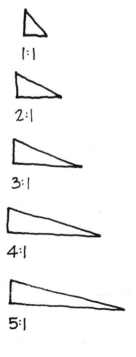

6-5. A comparison of various slope ratios.

Slope Orientation

Slope orientation significantly affects the site's climate, particularly in the areas of solar energy, solar reflection, precipitation, runoff, ice buildup, wind-scouring, and erosion. A general overview of orientation, done in conjunction with an analysis of the degree of slope, will indicate potential problem areas. South-facing slopes (in the Northern Hemisphere) receive the greatest amounts of solar radiation per year, which is desirable as long as the reflected heat does not exceed the building's heat-gain requirements by so much that mechanical cooling controls are needed. South-facing slopes are exposed to high light levels and fast evaporation of moisture for much of the year, resulting in challenging growing conditions for plant materials. Because these slopes warm up quickly during the winter months, they are less susceptible to ice buildup and therefore constitute a desirable orientation for locating safe pedestrian walking surfaces.

North-facing slopes allow ice buildup with little chance of melting during the winter months in the temperate and cold macroclimates, particularly if the building or plant materials mass immediately south of the slope is tall enough to block most of the sunlight from reaching the slope. Because north-facing slopes remain cool and shady from fall until spring, they do not experience as great a daily temperature fluctuation or as much evaporation or reflection as south-facing slopes during those seasons. A lower maximum slope is required in north exposures to allow wintertime access in the cold and temperate macroclimatic zones.

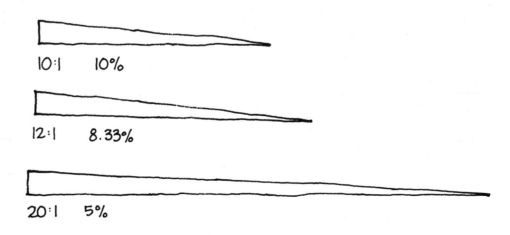

The actual allowable minimum and maximum slopes for a particular project depend on the intended use of the site, the type of maintenance available, soil conditions, and the project budget. The impact of different site uses on establishing appropriate slopes is obvious: a tennis court could not be built on a steep side slope, nor could a path planned for wheelchair access contain steps and still function as intended.

The degree and orientation of slope partially determine the cost of constructing and maintaining the project. Very flat and very steep sites can be expensive and require high maintenance. Flat slopes create various problems in draining the water away from desirable site features. The cost of installing an underground storm drainage system is often prohibitive because there is no existing system to adjoin.

On the other hand, steeper slopes may preclude mowing of grasses and forbs, may necessitate the use of a soil-stabilizing groundcover, or may require large-scale regrading and the use of retaining walls or terraces. In addition, certain soils are unstable at steep slopes and may dictate the use of a particular stabilization method over all others to reduce destructive erosion. Cut slopes (those created by cutting earth away and removing it from the existing site) can generally be held at steeper grades than fill slopes (those created by importation and addition of earth to site) can because no settling or compaction of the soil need occur with cut slopes to achieve stability.

Some sites slope uniformly, while others vary both in degree and orientation of slopes. A site composed of rolling hills will require a different type of design than one with a single, sled-run type of layout. The transition points between areas with different slopes and between sloped and flat areas should blend smoothly with one another to produce a natural, flowing landscape (figure 6-6).

Subsurface Drainage Features

The key question when analyzing subsurface conditions is whether there exists a subsurface storm drain system on the site or at a reasonable distance from it. If there is no such system, the designer will be forced to surface-drain the site. If a system is present, its size, condition, and vertical and horizontal location should be assessed.

The size of the existing system determines the (minimum diameter) size of new pipe that can be used and thus defines the drainage area that can be handled by each inlet (the smaller each drainage area is, the smaller the volume of water is that must be handled at each inlet, and the smaller the inlet pipe diameter

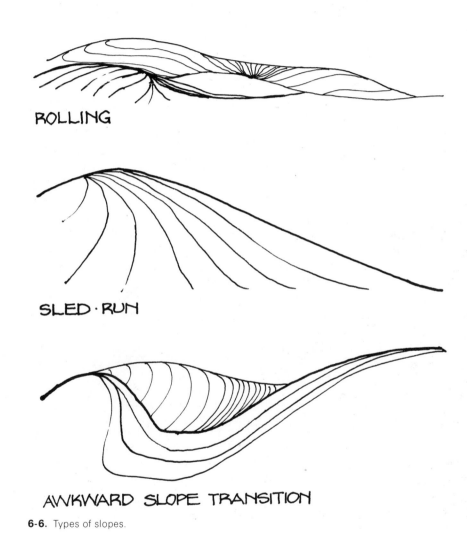

ROLLING

SLED·RUN

AWKWARD SLOPE TRANSITION

6-6. Types of slopes.

need be). The construction and condition of the existing system influences its capacity to handle runoff and helps decide what materials can be used in constructing the new system.

The overall design and project cost are also influenced by the system's location, which may require either the relocation of lines (often at considerable cost) or compromising on the location of certain site features. The system's horizontal location, or depth below grade, sets the depth of the new system and determines the surface grade (figure 6-7).

Soil and Water Conditions

Hardpan or excessively permeable subsoils, underground streams and old stream meanders, and high water tables may influence the outcome of the grading plan.

Hardpan soils are heavily compacted and impenetrable by water. Their presence on a site constitutes a barrier to drainage because water cannot percolate through them to be dispersed.

6-7. Installation of storm drain lines.

Excessively permeable subsoils create a different problem; water passes through them so quickly that soil nutrients leach out and plant materials are unable to thrive.

Underground streams and old streambeds create unstable surface conditions because of the constant shifting and high organic content of the resulting soil. The structure of these soils is usually not adequate to support construction without major alteration; attempting to add surface-drained water to these soils will only aggravate the problem.

A high water table again influences the condition of the surface above it, since precipitation and runoff have very little area through which to percolate and gradually disperse before reaching the saturated layers of subsoil. In addition, a high water table presents problems with the bearing capacity of the soil and with providing decent growing conditions for plant materials.

Off-Site Conditions

Similar assessments should be made of grading and drainage conditions on adjacent sites. It is particularly important to determine whether surface drainage from those sites flows directly toward, away from, or around the project area, and to identify with care the types of surfaces present on those sites.

GRADING DESIGN CONSTRAINTS

The designer rarely has unlimited freedom in establishing new grades on a site—especially if the site is small, confined, or filled with features that are to remain. Several factors act as design constraints or parameters.

Existing Mechanical Systems and Utilities

As was mentioned previously, the absence or presence of a subsurface drainage system on the site or on adjacent properties, the size and condition of the system, and its vertical and horizontal location all determine what can be done to drain the site by subsurface means.

Other underground utilities can create hardships as well. Buried telephone, electrical, gas, water, and sanitary sewer lines may all be found within the limits of the property. Their location may dictate the placement of above-grade site features so that costly relocation of the lines is avoided. Since underground lines are commonly placed at depths ranging from eighteen inches (or less) to ten feet (or more), prior knowledge of their location will keep the designer from exposing shallow lines, burying lines that require routine access too deeply, cutting through lines, or building or planting on top of them (figure 6-8).

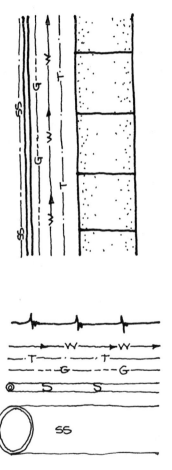

6-8. Utilities between the curb and sidewalk.

Existing Drainage Patterns

Water law in most states is quite specific in outlining what can and cannot be done to change existing, well-defined drainage patterns. The law essentially addresses increases and decreases in the quantity of water draining through an area and the path the water takes. Legal ramifications notwithstanding, many sites contain drainage patterns that, unless maintained in a condition similar to their natural state, will cause drainage problems on other parts of the site or on adjacent sites.

Quantity

By law, the amount of water entering or leaving a site may not be significantly increased (by diverting it around or holding it on the property) or decreased (by decreasing the permeability of the surface and thereby increasing the amount of runoff). Diversion of water entering the site from higher up the watershed by blocking it with walls, structures, or earth forms will change the natural course of its flow. If the reason behind this diversion is to decrease the surface flow across the property, either the water should be taken underground into a storm sewer system or the permeability of the surface it flows across should be increased to allow greater absorption.

Diversion or impoundment of the natural flow of water have been the source of several state and regional legal battles, beginning when a state through which the water feature passes attempted to impound the water with the result that the states on the receiving end of the watershed were prevented from receiving most of the normal benefit from it. Impoundment, whether at the scale of a dam on a major river or at the scale of excavating a stock watering pond, changes the drainage characteristics, growing conditions, and general character of the land, both above and below the impoundment.

Such developments as shopping centers, industrial parks, and high-density residential areas have a much greater proportion of paved or hard-surfaced areas than undeveloped sites. The consequent reduction in permeability causes more surface runoff onto adjacent properties. When these types of developments are planned, the designer must take steps either to contain the additional runoff on the site or to channel it underground to a storm sewer system.

Location

The drainage patterns across a site can be changed dramatically, as long as the locations of the source and discharge points remain unchanged relative to each other. Reversing the direction of downhill flow is both extremely difficult and very costly, so the overall natural drainage pattern on a site after grading is completed usually remains similar to the original pattern. However, on small or constricted sites, a change in the degree or direction of slope can produce a change in the flow onto or from adjacent property.

Accessibility

Degree and orientation of slope, discussed earlier, affect who has access to a site, when, and how easily. A desire for maximum accessibility is especially significant in guiding the planning of access to steep sites because of the long horizontal distance needed to make the rise manageable. How the site is to be serviced also affects the site grading. Tolerance for degree of slope in an approach varies greatly between vehicles: some can negotiate a short, steep slope, while others must have gradual inclines. The general layout of the circulation pattern should be considered during formulation of the initial grading design, because the circulation plan must be compatible with the grading profile to ensure that slopes that may work on a straightaway but not on a turn do not appear in the plan (figure 6-9).

Special Uses

A program may call for special site use areas, such as a dining terrace, a jogging trail, a playground, or a sports field. Each of these uses demands a certain type of grading plan to function properly; providing adequately for the special functions may entail developing the rest of the site grading plan around the special use areas.

Site Features to Remain

The new grading plan must accommodate the site features that are to remain. These features may include existing buildings with established entrance levels and lower-floor windows or docks; buildings constructed from materials that cannot be exposed or buried; streets, sidewalks, and other paved areas; storm sewer lines and other utilities; trees and shrubs; and natural features such as rock outcroppings.

Existing Buildings

The entrances and other openings into a building obviously must be kept clear of earth. Often, though, the materials of which the building is constructed will limit the extent to which earth can be built up or removed. Some materials (notably wood and porous masonry) deteriorate rapidly when left in contact with soil moisture or are subject to infestation by destructive insects such as termites when exposed. Exposing more of the foundation than showed previously may also be undesirable from an energy conservation standpoint. Maintaining positive drainage is a critical step in avoiding damage to any existing structure.

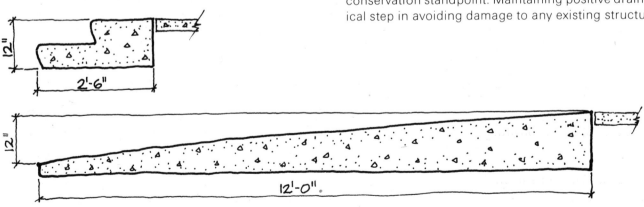

6-9. Comparison of the horizontal distances needed for steps and for a ramp for accessibility.

Streets, Sidewalks, and Parking Areas

Not only must the proposed elevations of streets, sidewalks, and drives match the elevations of existing ones that are to remain, but the transitions between existing and proposed surfaces must be smooth and gradual, causing neither a ponding spot nor a place where bicycles and cars will "bottom out." Curb heights should match existing levels and should be high enough to form a definite, visible step—but not so high as to cause car doors to scrape.

Trees and Shrubs

The amount of earth disturbance that trees and shrubs can tolerate is widely misunderstood. Far too many designers think that if they leave a "well" around the trunk of the tree (usually of a diameter less than five feet across), the tree will survive because it has not been buried—regardless of the quantity of additional earth placed outside the zone of the well. While it is true that such treatment will avoid causing the tree to die from having its bark covered with soil, it is quite likely that the tree will suffocate because the tree roots (previously a certain distance under the surface) will no longer have the same access to air and water as a result of the increased earth cover outside the well radius.

The roots providing a tree with its nourishment and structurally supporting its canopy extend beyond the dripline (the outermost reach of its branches). To be effective, then, a tree well should extend at least to the dripline, and everything within the dripline should remain undisturbed. When variation in elevation within the dripline is required, it should be as slight as possible. Although some species of vegetation tolerate adjustments in grade more easily than others do, a general rule of thumb is that no more than six inches of depth should be removed from within the dripline and no more than twelve inches of depth added (figure 6-10).

Another point to consider when designing a site to preserve existing plants is the impact of construction on the soil supporting those plants. Even if the grade beneath the plants is not disturbed by cutting and filling, excessive soil compaction during grading or construction operations can cause decline and even death of the affected plants. Care should be taken to establish an access route for construction vehicles that will not force the trucks to drive under the driplines of the trees that are to remain. The compacting action of large vehicles—particularly when it occurs on wet soils—forces the air out of the spaces between particles of soil and smothers the tree's roots. As one part of providing construction protection, construction fencing should be installed at the dripline of all plants that are to remain.

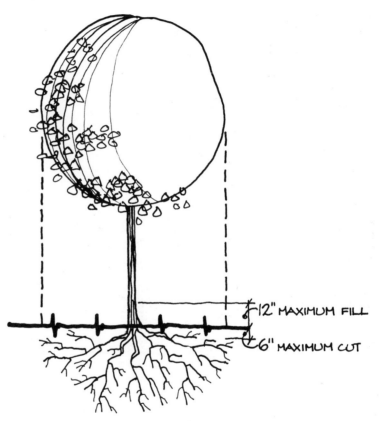

6-10. Grading should be outside the dripline.

12" MAXIMUM FILL

6" MAXIMUM CUT

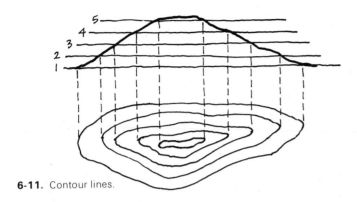

6-11. Contour lines.

DEFINITIONS AND TERMS

Familiarity with terms is a must for anyone working with grading and drainage. Following are a few commonly used grading terms and their definitions.

Contour Line: A line connecting all points of equal elevation above a given datum (figure 6-11).

Datum: A reference plane used in surveying for the establishment of a starting elevation. Sea level is the datum used by the United States Geological Survey (USGS).

Benchmark: The known datum that is used to establish a site survey. This may be an existing benchmark within the USGS system, or an arbitrary elevation set to aid the contractor in site construction.

Contour Interval: The vertical distance between contours. Common intervals are one, two, five, and ten feet. A plan drawn with a two-foot contour interval, therefore, would show the incremental changes in elevation (contour lines) every two feet (figure 6-12).

Longitudinal Slope or *Grade*: The amount of slope in the long dimension of a plan element (figure 6-13).

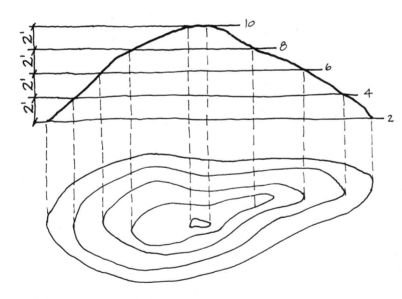

6-12. Contour interval of two.

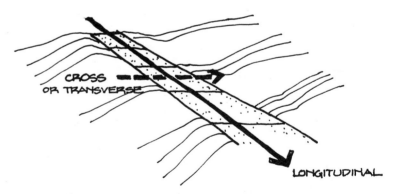

6-13. Longitudinal slope and cross slope.

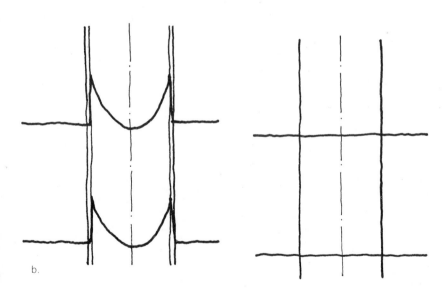

6-14. Travel distance; a. curb and crown; b. flat road.

Cross Slope or *Transverse Slope*: The slope occurring at right angles to the longitudinal slope of a plan element (that is, in the crosswise direction) (figure 6-13).

Rate of Slope: The ratio of the horizontal distance (run) to the vertical distance (rise), such as 10:1, 2:1.

Travel Distance: The horizontal deviation from a straight line that a contour will follow when going up or down a road crown or a curb. When a contour crosses a road that has neither crown nor curb, it continues in a straight line across the road's surface. If, however, the road is crowned or swaled, the contour is forced to travel back up the hill (to traverse a swale or to go down a curb) or to travel back down the hill (to surmount a crown or to go up a curb) in order for the contour line to maintain a path of equal elevation (figure 6-14).

Crown: The rise to a high point in a road or other man-made surface; also, the top of a hill.

Ridge: The highest elevation of a hill—V-shaped in plan execution.

Swale: A drainage path on a site, shown by an upside down V in plan execution.

Spot Elevation: An elevation given for the purpose of clarifying the grade in a particular location where more detail is required than is provided by the contour lines.

Station Points: The points of reference used by a surveyor in establishing elevations, usually established along a linear site element such as a road. The point of beginning, or the first elevation taken, is always designated as Station 0+00, and stations are always located 100 feet apart. Any odd distance between two stations is referred to as a "plus station," and added to the preceding station. The linear distance of 54 feet, for example, is referred to as Station 0+54, and the linear distance of 154 feet is referred to as Station 1+54 (figure 6-15).

Finish Floor Elevation: The elevation of the finish floor of a structure. This *must* be shown on the site plan, with a finish floor given for each change in elevation (such as from the living area to a lower walk-out level). The finish floor elevation is given two ways on all drawings: as an architect's finish floor, which is most commonly designated as 100.00, and as an engineer's elevation, which will correspond to the contour of the property, and may be something like 43.75.

Batter: The amount of deviation from a true vertical line, used in reference to the faces of retaining walls that are sloped back against the hill at the rate of a fraction of an inch or so per foot.

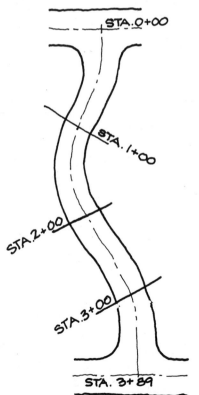

6-15. Station points.

THE GRADING PROCESS

Although trial and error is usually involved in the design of a grading plan, it can be kept to a minimum if a logical grading process is followed.

Topographic Survey

First, the designer should become acquainted with basic information about the existing topography of the site. This is most often obtained from a topographic survey, done by a surveyor hired by the owner, with the designer providing the surveyor with a list of information needed. Elevations are commonly given for all of the following: high and low points; building corners and other structures; tops and bottoms of curbs, walls, steps, and ramps; corners of paved areas and terraces; intersections of paved areas; manholes, grates, valves, and inlets; inverts (flow lines) of all storm sewers; and trees with trunks over a certain diameter. By giving the elevations of such features as trees, the surveyor is also locating them on the site, although not as accurately as is done on the property survey (see Chapter one).

The topographic survey is based on information gathered from a series of spot elevations. These may be taken randomly (at high and low spots and other landforms indicative of the character of the land, as apparent to the naked eye) or by means of a grid system. The grid may be laid out as small as a measurement every ten feet or as large as a measurement every one hundred feet or more, depending on the scale of the project, the variation in topography (the more variation exists, the greater the number of shots are needed for accuracy), and the degree of plan accuracy required. Usually, a surveyor using a grid system will also give the elevations of obvious site features that do not occur exactly at a grid point, thus combining the best features of both the random and grid survey systems. The actual document used by the designer in developing the site plan may in fact be a combination property-topography survey.

The survey may be presented to the designer in one of three forms. First, it may show the actual shots (the plus and minus shots) as recorded in the field book. Second, it may show the elevation of each shot converted to feet above sea level. Third, it may provide a plot of the contours. The last is the most useful and the most expensive to obtain; but because the contour lines are needed for developing a site grading plan, the designer will have to locate them if they are not on the survey. This is done by interpolating between spot elevations. Even if contour lines are shown, interpolation may be needed to clarify the grade around a particular site feature.

Interpolation

Interpolation is the calculation of an unknown elevation based on the known elevations of two other points, one on either side of it. It is used not only to find the whole contour lines between spot elevations, but also for such things as finding the corner elevation of a building set into a constantly sloping hill, determining the elevation of a retaining wall corner, and finding the top or bottom of a curb. There are two methods of interpolation commonly used in plotting site contours: graphic and mathematical.

Graphic Interpolation

Graphic interpolation relies on the use of equal divisions to determine the distances between two points or contour lines. It is quicker than mathematical interpolation, but not as accurate. However, because of the relatively large margin of error acceptable in grading work (and for large-scale projects and schematic design), the precise location of an elevation on a plan is often not necessary.

The following method is used to interpolate graphically between spot elevations (figure 6-16):

1. Determine what, if any, whole-number contours will occur between the two points. For the example shown in figure 6-16, contours 87, 88, and 89 occur between spot elevations 86.62 and 89.81. If the spot elevations were 86.62 and 86.81, however, no whole-number contours would fall between them, and interpolation would not be necessary.

2. Drop the decimal points. Graphic interpolation uses units of whole equal measure; thus, 86.62 becomes 8,662 units. The two points of known elevation must be given the same degree of accuracy initially—either both in tenths or both in hundredths. Otherwise, dropping the decimal point would create a large error in the actual number of units represented by each spot elevation. For example, if elevation A were 86.6 (instead of 86.62) and elevation B were 89.81, the failure to add a zero to 86.6 in order to bring points A and B to the same unit of accuracy would lead to a comparison (after decimals were dropped) of 866 units with 8,918 units.

3. Find the mathematical difference between points A and B.

4. Draw a straight line between points A and B, and then a line perpendicular to that from point B.

5. Place the "zero" end of a scale or other instrument marked with equal divisions on point A and swing the other end at an angle until the 319 mark (the difference between 8,662 and 8,981) crosses the perpendicular line.

6. Determine the mathematical difference between 8,662 and the first whole-number contour (87.00 or 8,700). This difference is 38 units.

7. Locate the 38 mark on the scale and drop a perpendicular from this point to the line between A and B. This is the interpolated point at which the 87 contour occurs between points A and B.

8. Repeat the process for the remaining whole-number contours between points A and B.

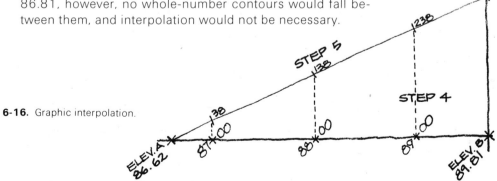

6-16. Graphic interpolation.

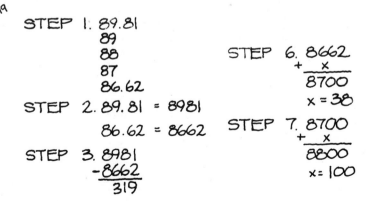

STEP 1. 89.81
89
88
87
86.62

STEP 2. 89.81 = 8981
86.62 = 8662

STEP 3. 8981
−8662
_____
319

STEP 6. 8662
+ x
_____
8700
x = 38

STEP 7. 8700
+ x
_____
8800
x = 100

The process also works in reverse, beginning at point B and going toward point A. This is a good way to check accuracy.

To be as accurate as possible in plotting contour lines, the designer must take care to interpolate between every two points wherever a whole-number contour line will occur. This means interpolating the diagonals on a survey grid as well as the perimeters of the squares because a site feature such as a narrow swale could be hidden between the grids.

Once the spot elevations representing whole-number contour lines have been located, the spots are connected and the contour lines drawn.

Mathematical Interpolation

Mathematical interpolation utilizes some of the same steps as graphic interpolation, but is based on the proportional relationship between the difference in elevation and the measured horizontal distance between the two points. Using the same two points A and B, the contours between them would be mathematically interpolated as follows (figure 6-17):

1. Find the total difference in elevation between the two points, in accordance with Step 2 of the graphic interpolation method.
2. Measure the distance between the points, using the same scale as that at which the drawing is made. Point A scales 196 feet from point B.
3. Find the mathematical difference between point A (86.62) and the first whole contour (87). This figure is 0.38 feet.
4. Set up a proportion comparing the ratio of elevations (point A to first contour/point A to point B) to the ratio of horizontal distances between points (same): 0.38/3.19 = ×/196. Solving this equation gives a value for × (the linear distance from point A to the first contour) of 23.34 feet.
5. The difference between point A and contour 88 is 1.38. Again set up a proportion of elevations to linear distances—this time, 1.38/3.19 = ×/196.
6. Continue this process until all whole-number contours have been plotted. As is the case with graphic interpolation, the process may be reversed.[1]

READING AND INTERPRETING CONTOURS

The finished contour map is of no value if the designer does not know how to interpret what it says. A clear understanding of what the contour map is—a plan representation of a three-dimensional earth form—will help the designer develop a grading plan quickly. Following are some specific facts to help in understanding how to read contours.

The spacing between contours indicates the severity and uniformity of the slope. The closer together the contour lines are, the steeper the slope is. Widely spaced contour lines indicate very gradual slopes or flat land. Contours spaced at equal distances from one another indicate a convex slope or a hump. Contours that are widely spaced at the bottom of a hill and tighten up as elevation increases indicate a concave slope or a hollow. Contours that widen and tighten across a site indicate a series of undulating hills (figures 6-18 and 6-19).

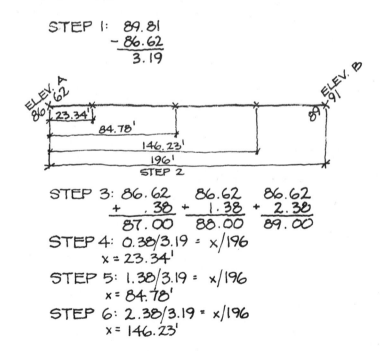

6-17. Mathematical interpolation.

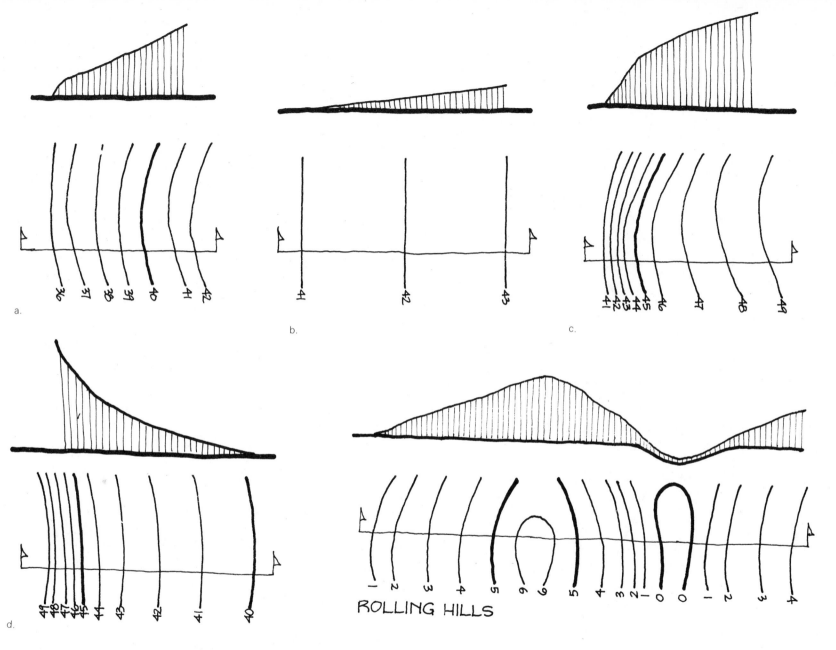

a.

b.

c.

d.

ROLLING HILLS

6-18. Maps: a. even; b. convex; c. concave; d. irregular.

6-19. Rolling slope.

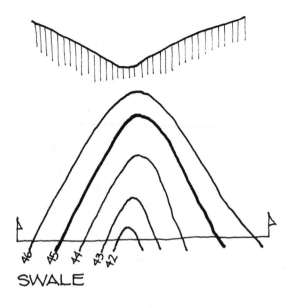

SWALE

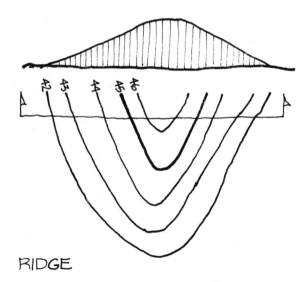

RIDGE

6-20. Swale and ridge contours.

V-shaped contours indicate a ridge or a valley. Since water flows perpendicularly to the point of greatest slope, the V shape sheds water either to its center (the valley) or to its sides (the ridge). The two landforms can be differentiated in plan because the point of V is situated uphill of the arms (against the flow of the water) for a swale and downhill of the arms (with the flow of the water) for a ridge. If a spot elevation is given on a V-shaped feature, it will be a high point on a ridge and a low point or flow line on a swale. The narrower the distance between the arms of the V (and the sharper the point), the steeper the landform is (figure 6-20).

Every contour is a continuous line that forms a closed figure over some finite area on the surface of the earth. This usually does not occur within the boundaries of the site, but contours must close on themselves if three-dimensional earth forms are to exist. For example, if an elevation midway up a hill did not occur on both sides of a hill, (or in other words, if a closing contour did not exist) it could only be because the "hill" never again fell to the medium height identified on its front side (figure 6-21).

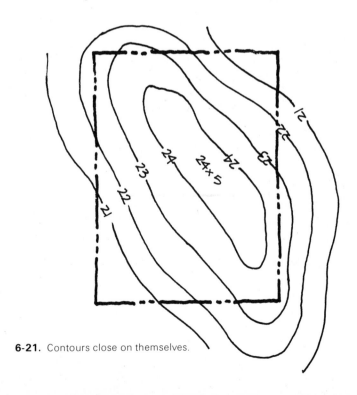

6-21. Contours close on themselves.

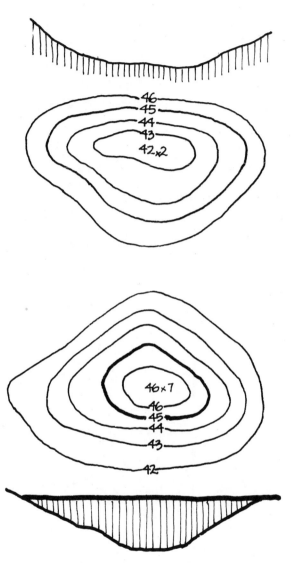

6-22. Depression and mound contours.

Closed contours surrounded by other closed contours indicate either a depression or a mound, with the difference being shown by the progressive numbering of the interior contours and by the spot elevation given at the center. When closed contours occur within the confines of a site, the designer (barring economic constraints) can regrade the site to eliminate the contours entirely, since they need not be matched at a property line. Closed contours are eliminated by filling low spots or leveling high ones (figure 6-22).

Contours always occur in pairs. This is necessary to enable them to define elevations of three-dimensional landforms. A contour represented as a single line (which cannot by itself enclose an area) shows a knife-edge of equal elevations occurring exactly on an even contour number—a certain misrepresentation (figure 6-23).

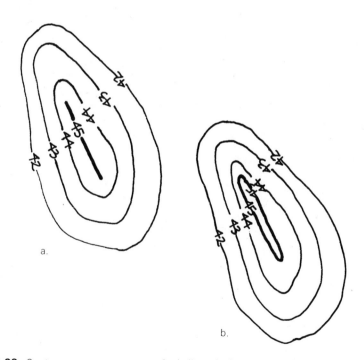

6-23. Contours: a. never occur as single lines; b. but always as pairs.

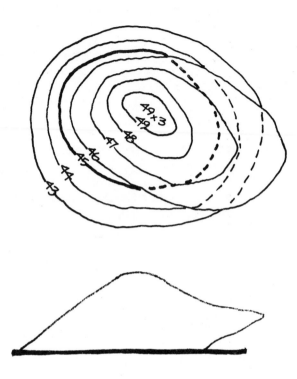

Contours never cross one another, except in the condition of an overhanging cliff. Were two contours to be shown as crossing in any other situation, it would indicate the existence of two different elevations simultaneously for the same point. Although it appears that contours on the vertical face of a building or retaining wall are stacked on top of one another, a perspective shows that these contours actually wrap the vertical face of the structure or excavation in very distinct striations (figure 6-24).

Contour lines run perpendicular to the steepest slopes. Since contours represent equal differences in elevation between one line and the next (each point on contour 47, for example, is a foot higher than each point on contour 46), it stands to reason that the farther apart the two lines are, the shallower the slope must be. Water naturally follows the steepest slopes, running perpendicular to the contour lines.

Contours must match existing perimeter grades by the time they reach the property line. Were this not to happen, a cut or fill situation would exist on the property line, damaging neighboring property (figure 6-25).[2]

6-24. Contours cross *only* for overhanging cliffs.

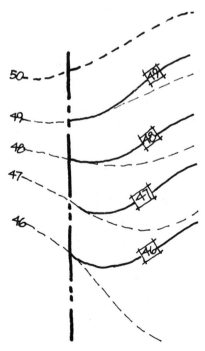

6-25. Contours must match existing grades at the property lines.

Setting Slopes

With a general picture of the site's character in mind, the designer can begin setting slopes for the proposed grading plan. Slopes can be set randomly to form interesting and pleasing patterns on the land, and often they are—in open spaces and in locations where a defined slope is not needed to serve a particular function. In many other instances, however, a specific degree of slope or range of slopes is needed.

To calculate slope, two points of known elevation and the distance between them are needed. Both the percent of an existing slope (100%=vertical) and the degree of a desired slope can be calculated using the slope formula, which compares rise (change in elevation) to run (distance between points). The slope formula is:

$G = d/L$

where:

G = Gradient or slope (the answer is always multiplied by 100 to give an answer in percent)
d = Mathematical difference in elevation between two points of known elevation (rise)
L = Horizontal length between two points of known elevation (run)

For example, given points A and B 78 feet apart, of elevations 67.52 and 69.19 respectively, calculate the slope:

1. Find the difference in elevation between points A and B: This is equal to 1.67.
2. Since L was given, plug the numbers into the slope formula:

$G = 1.67/78 [\times 100]$
$G = 0.0214 \times 100 = 2.14\%$ slope

Any variable in the formula can be determined, given fixed values for the others. On the basis of a set percent of slope and two points of elevation, a designer can determine the length between those points. Knowing the elevation of one point, the length between points, and the slope, a designer can find the elevation of the other point.

If it is necessary to construct a site feature such as a parking lot at a constant slope, and if the desired percent of slope, the length of the lot, and an elevation at one of the ends of the lot are known, the elevation at the other end can be calculated (figure 6-26).

For example, assume a constant slope of 3.33%, a parking lot length of 125 feet, and an elevation at the high end of the lot of 57.24. The elevation at the low end of the lot can be found as follows: a 3.33% slope carried over 125 feet would amount to a total change in elevation of 4.16 feet (0.0333 × 125); 4.16 feet subtracted from the high-end elevation of 57.24 feet leaves 53.08 feet—the low-end elevation.

A second formula can be used to determine how far apart contours should be located on a plan. This formula is:

$D = CI/G \times 100$

where:

D = Horizontal distance between contours (the scaled distance on a plan)
CI = Contour interval
G = Gradient or slope (from the preceding formula)

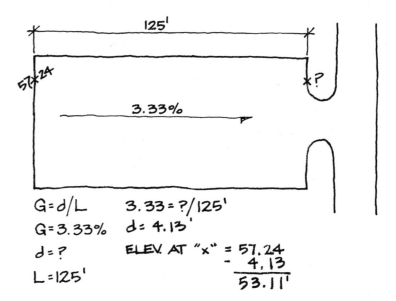

$G = d/L \qquad 3.33 = ?/125'$
$G = 3.33\% \qquad d = 4.13'$
$d = ? \qquad \qquad$ ELEV. AT "x" = 57.24
$\qquad \qquad \qquad \qquad \qquad \qquad - \quad 4.13$
$L = 125' \qquad \qquad \qquad \qquad \qquad 53.11'$

6-26. Finding an elevation using the slope formula.

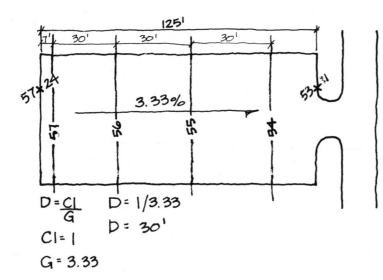

$$D = \frac{CI}{G}$$
$$D = 1/3.33$$
$$D = 30'$$
$$CI = 1$$
$$G = 3.33$$

6-27. Finding the measured distance between contours at a given slope.

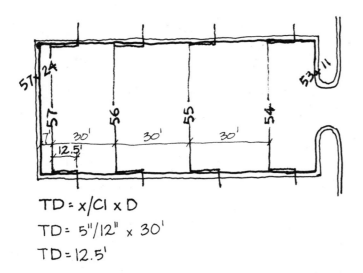

$$TD = x/CI \times D$$
$$TD = 5''/12'' \times 30'$$
$$TD = 12.5'$$

6-28. Finding the travel distance.

The use of this formula is dependent on the gradient formula just given. The horizontal distance between contours varies as the slope varies: the steeper the slope is, the closer together the contours will be on the plan.

Working from the parking lot problem given in figure 6-26, assume that the next step is to plot the whole-number contours at one-foot intervals (figure 6-27).

Using the slope of 3.33% given, plug the variables into the formula:

$$D = 1/3.33 \times 100$$
$$D = 30 \text{ feet}$$

Beginning at the high end of the lot, the first contour to plot will be 57. Subtract 57 from 57.24. To plot this contour, use the slope formula to find the length between two points of known elevation (57.24 and 57):

$$G = d/L$$
$$3.33\% = 0.24/L$$
$$L = 0.24/0.0333$$
$$L = 7.27$$

Contour 57 is therefore roughly 7 feet away from spot elevation 57.24.

A third refinement of the slope formula is used to calculate how far a plotted contour travels along the face of a curb, ramp, or other vertical deviation from the path the contour would take if the surface remained level.

Travel distance (TD) is the uphill or downhill deviation of a contour from the straight-line route caused by a vertical rise or drop along the direct route. It relates the contour interval to the vertical rise (or drop) and multiplies this number by D (the distance between contours at a given slope), determined from the formula above.

For the same parking lot problem, assume a 5-inch-high curb along all sides of the lot. A common error in using the travel distance formula is to attempt to relate the vertical difference in elevation, (which is usually given in inches) to the contour interval (which is usually given in feet), without converting them to the same measure. The travel distance formula is:

$$TD = x/CI \times D$$

where:

x = Vertical rise or drop
CI = Contour interval
D = Distance between contours at a given slope

Convert the vertical rise and the contour interval to the same measure. This gives an x value of 5 inches, and a CI value of 12 inches. Then, since D is already known, plug the variables into the formula:

$$TD = 5''/12'' \times 30'$$
$$TD = 12.5'$$

The deviation from the straight-line route across the road is therefore 12.5 feet each way.

This means that, instead of traveling straight across the parking lot (as they would if there were no curb), the contour lines will travel back up the hill as they make their way down the curb into the road from one side and will travel down the hill as they make their way up the curb out of the road on the other side (figure 6-28). The "downhill-to-go-up" and "uphill-to-go-down" rule at first seems odd when described in plan view, but it makes sense when considered in connection with the fact that contours describe equal elevation. To attempt to go down the hill to get down the curb would require the contour to be placed in thin air as soon as it left the curb, which is contrary to the nature of contours as lines moving at constant elevation along the land's surface. If seen along the curb's face, the climbing-to-go-down contour line would describe a simple diagonal.

Calculating travel distance can be more complex than accounting for one simple curb jump. A crowned road with a transversely sloped sidewalk adjacent to the curb would have three possible different travel distances: one for the transverse slope on the walk, one for the curb, and one for the crown.

CREATING POSITIVE DRAINAGE

Whether the grading plan calls for something as simple as minor adjustments in the existing spot elevations or as complex as moving thousands of tons of earth and carefully calculating each slope using the above formulas, one unbreakable rule applies: the surface drainage must run away from site features that are to be preserved. This drainage can be handled through a combination of sheet and swale drainage, but all water is eventually carried off-site in a swale.

The swale may be so shallow as to be nearly imperceptible to the casual observer, or it may be well defined and moatlike. Determining which contour to use as the swale line is one of the first trial decisions to be made in developing the grading plan.

THE PROCESS OF GRADING A SITE

The first step in grading a site is to establish grades at all existing site features that are to remain, in accordance with the analysis. Since these are permanent features, all proposed grades must be shaped to match the grades on which the features are situated; it is easier, therefore, to start with these existing known grades and work toward the areas that are open to alteration rather than trying to hit the fixed points through trial and error.

The second grades to be set are those for features that must be smooth and constant (and usually must lie within fixed maximum and minimum limits): parking lots, roads and walks, playing fields, outdoor terraces. There is more room for compromise in setting these grades than in working with the fixed site features. If variation in the percent of grade of a certain site feature is possible, the designer may wish to grade the feature at two or three alternate percents, to establish the absolute high and low elevations.

The finish floor elevations of all structures can then be assigned, based both on the existing grades (to avoid excessive cut and fill) and on the desired approach slopes (for access and aesthetics). Usually, the finish floor elevations can be used to take up the slack in the grading of all other site features, varying from a few inches to several feet. Finish floor elevations are given for all levels and level changes at or below grade on the grading plan, whether these changes occur in separate structures or on separate floors of the same structure.

It is extremely important that the finish floor elevations be given in terms of both engineering and architectural scales. Engineering elevations are based on the benchmarks and survey elevations used for the project. Architectural elevations are usually based on a first-floor elevation with an assigned value of 100.00. Failure to state both the engineering and architectural elevations can lead to mistakes in construction, especially if the engineering elevations should happen to be close to 100.

For example, assume that the desired finish floor elevation for a given project is 1,198.50 engineering, to correspond with 100.00 architectural. It is customary to drop all numerical places larger than the tens, making the notation for the engineering elevation 98.50. If the designer failed to specify that the architectural elevation of 100.00 was equal to the engineering elevation of 98.50, the contractor might well assume that the finish floor elevation was to be 100.00 engineering and might place the structure eighteen inches higher than the designer intended it to be, inviting all sorts of grading problems in other areas of the site.

When a trial finish floor elevation has been set, the designer determines the desired slope away from the structure or other site features and sets the first contour. This contour is the most important one on the site because it is the one that ensures positive drainage. The first contour away from the finish floor elevation is always the first whole number in the contour interval sequence lower than the finish floor elevation. For example, if the finish floor elevation is 87.5 (and if the contour interval is one), the first contour shown around the structure will be 87. If the contour interval is two instead of one, the first contour shown will be 86 instead of 87; and if the contour interval is five, the first contour shown will be 85.

It should be obvious from the two- and five-foot contour interval examples that considerable differences in design could be hidden if the only elevations given at the face of the structure were the finish floor elevation and the next whole contour in the contour interval sequence. A jump from 87.5 to 86 could indicate either a constant slope at a two-foot interval, or a step at the threshold. To avoid ambiguity, a spot elevation should be placed at all building entrances.

Depending on soil conditions that determine runoff, aesthetics, cost, and available space, the first whole contour below the finish floor elevation may or may not encircle the site feature. If it

does not, a spot elevation on the uphill side of the structure will be necessary to indicate that the water runs around the structure, not directly into it. This spot should be higher than the swale contour, but lower than the first uphill contour (figure 6-29a).

Setting the swale line and the desired slope from the site feature to that line is the next step after setting the trial finish floor elevation(s) and first contour. Since a swale forms an uphill-pointing V, the swale contour will be paired around the sides of the site feature—at least for a small distance—and will return to a single contour on the downhill side of the feature. There are no functional exceptions to this; attempting to point the V of the swale downhill rather than up and still accomplish positive surface drainage is doomed to failure.

A simple check that involves looking at the swale contour and the next contour above it can be performed to determine if the swale is placed correctly. Assuming that the lines are numbered correctly, a spot elevation between the swale contour and the next highest contour should be higher than the swale, but lower than the hill contour. However, since a swale drains to the center, the spot elevation between the paired contours of the swale must be lower than the swale contour. In figure 6-29b, for example, the spot elevation would have to be less than 25; yet the next highest contour above the 25 swale contour is 26, and there must be a 25 contour between spot elevation 24.5 and contour 26. Therefore, the swale is pointing in the wrong direction. The 24.5 spot does occur on the downhill side of the 25 swale contour, as it should—but downhill here means toward the very feature for which the designer is attempting to create positive drainage.

Although the swale, as a drainage channel, has been discussed primarily in relation to a site structure, it can also serve this purpose in some measure with respect to all site features. It is not necessary to place all features on distinct plateaus surrounded by well-defined swales (figure 6-29c); in fact, in many cases (such as across major parking facilities), the swale may be so gradual that the impression it gives is one of total sheet drainage. Terraces, landforms, recreational facilities, and roadsides all require positive drainage in some form.

Once the swale contours around all site features have been established, the remaining contours are filled in and connected with existing contours. Another check on the accuracy of the grading plan is to follow the contour numbers in numerical sequence: if a number is missing, either the numbers themselves are wrong or the designer has forgotten a contour. In no case can a contour be skipped, whether it should fall between two other contour lines or between a contour line and a spot elevation. In plotting contours on the site plan, all whole-number contours between other numbers are shown, just as when they are located through interpolation.

If at this point (or earlier in the grading schematic) the first trial finish floor elevations prove too high or too low, resulting in steep slopes or low spots, the finish floor elevations can usually be adjusted without changing the general pattern established.

When trial finish floor elevations are being set, the designer should also set trial high and low elevations for berms, mounds,

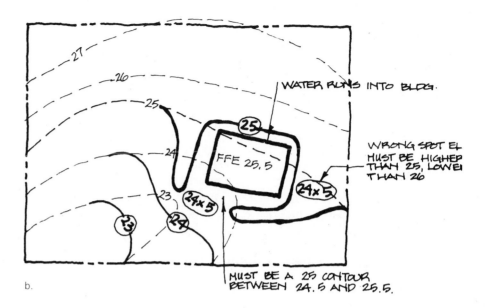

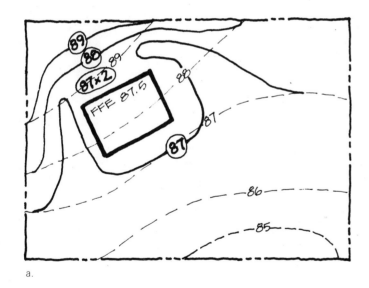

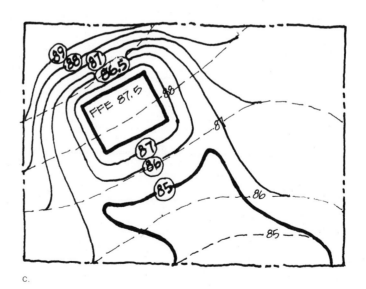

6-29. Locating swales: a. correctly placed swale showing spot elevation; b. incorrectly placed swale; c. swale around a building on a plateau.

walls, stairs, and ramps. Doing this at the beginning of the grading design process ensures that a clearer mental image of the site's character will develop, and the maximum and minimum heights of all site features can be adjusted simultaneously. To attempt to finish-grade a site piecemeal is to invite disaster in the form of constant reworking, grades that do not match, and awkward or abrupt changes and transitions between site areas.

From this point on, the grading process is a matter of continually adjusting and readjusting the contour lines (preferably only slightly) and the spot elevations. Refinements are made to accommodate steps, curbs, and other vertical changes, to improve matches between the architectural and natural profiles of the site, to limit or increase exposure (for windows and door openings) of large amounts of foundation, and to balance cut and fill for economy's sake.

Graphic Display of Contour Lines

Finished contours and spot elevations are shown on the proposed grading plan as heavy solid lines with large numbers that are circled or otherwise emphasized, distinguishing them from the lighter, dashed existing contour lines. Proposed spot elevations are also circled. Every fifth contour (at whole numbers divisible by five) is made heavier yet, to make the plan easier to read. Numbers are placed either on the uphill side of the contour

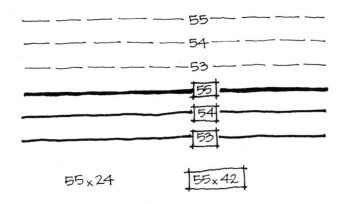

6-30. Graphic portrayal of contours.

line or (preferably) within the line itself. Numbers should be shown consecutively and with enough frequency that the contractor will have no trouble determining the elevation of a given line. Contour numbers should read right-side-up when the plan is turned so the base of the line of elevation at which the number appears is exactly perpendicular to the uphill slope at the same place. This means that many numbers will appear upside-down from any one map position, but once the reader becomes accustomed to this method of graphics, it becomes a simple way of telling whether a slope is rising or descending (figure 6-30).

Accuracy of the Grading Plan

The accuracy of a grading plan depends on the skill with which the survey was handled and the care that was taken in interpolating between spot elevations to locate the whole contours. Plans will never actually be graded on site to exactly the tenth or hundredth of a foot shown on paper because too many variables intervene: extraneous site conditions, the type of equipment used, and the skill of the operator. Actual grading that comes within a tenth of a foot of a specified spot elevation or contour line is extremely accurate—particularly since actual grades deviating by as much as a half-contour from those shown on the plans are legally allowable.[3]

CUT AND FILL

It is usually desirable to balance the cut and fill on a site, to avoid the cost and bother of either having to purchase suitable soil (at prices in excess of $15 per cubic yard) or to haul large quantities away. There are many different ways of calculating cut and fill, some of which are described below; however, a visual inspection coupled with a thorough understanding of how contour lines and grading operate will often tell the designer, almost as accurately as a series of complicated calculations would, whether there is a balance, a shortage, or an overrun of available soil.

Compensating for Settling

The accuracy of any soil's quantity takeoff depends to a great extent on the composition of the soil itself. All soils will settle after being disturbed and will become compacted when equipment is intentionally or unintentionally run over them. Failure to

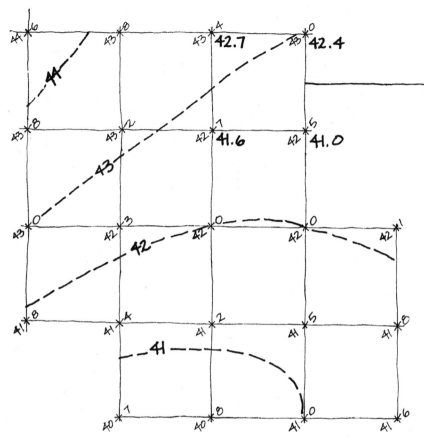

6-31. Grid point method.

$$
\begin{array}{cccc}
43.4 & 43.0 & 42.7 & 42.5 \\
-42.7 & -42.4 & -41.6 & -41.0 \\
\hline
0.7 & 0.6 & 1.1 & 1.5
\end{array}
$$

$$
\begin{array}{l}
0.7 \\
+\,0.6 \\
+\,1.1 \\
+\,1.5 \\
\hline
+\,3.9
\end{array}
$$

$3.9 \div 4 = 0.93$ FEET OF CUT (AVERAGE FOR THIS GRID SQUARE)

Methods of Calculating Cut and Fill

Three methods of calculating cut and fill are discussed briefly in this section: the grid-point method, the overlay technique, and the average end area.

Grid-Point Calculation

The grid-point method of calculating cut and fill is based on division of the site area into grid squares of equal area. Both the existing and proposed elevations at the four corners of each square on the grid must be calculated or estimated. The total volume of each grid square is calculated by averaging the difference in elevation at each of the four corners. This average is either greater or less than the average for the whole grid and constitutes, therefore, either a positive (fill) or negative (cut) number. (The total figure for the grid is determined by averaging the cut or fill figures for all individual grid squares for the entire site.)

The process can be simplified by combining the calculations for elevation differences (the difference between existing and proposed elevations at any given grid corner) that are common to more than one grid corner or, better yet, by using a computer. Since an adjustment of even a tenth of a foot over an entire site can make the difference between a balance and an excess or deficiency of soil, a computer can quickly and accurately apply a tenth-of-a-foot difference (or more, if needed) to all numbers on the grid. A simple application of the grid point method of calculating cut and fill is used to determine the amount of earth to be removed from excavations (figure 6-31).

include sufficient soil to compensate for shrinkage and settling will eventually result in additional settling of the soil in disturbed areas, which can reverse the established drainage pattern and run water into a building or can cause ponding in relatively flat areas, making them unusable. Predictions of how much a particular soil type will settle under an assumed set of conditions are estimates at best, which introduces into the carefully prepared grading plan an immediate degree of potential error. Compaction factors can range from near zero (for wet sand) to more than one-third of the total volume of soil.

The Overlay Technique

The overlay technique produces a much rougher estimate and is often used in the initial stages of the grading process to get a general idea about whether or not the site is balanced. It involves marking on a plan all points of zero change (that is, all points where the existing and proposed elevations are the same) and then outlining areas of cut on one overlay and areas of fill on another.

The areas of cut and fill can be more quickly and accurately determined by using a planimeter—an instrument consisting of two thin arms joined by a ball-and-socket, with a weight on one end to hold the instrument firmly on the page and a tracer needle and magnifying glass on the other. As the designer moves the tracer arm around an area's outline, a series of discs revolve and record the area in square inches. Conversion factors enable the designer to relate the square-inch figure to the accuracy of the instrument and the scale of the drawing, and the actual cut and fill areas can then be determined (figure 6-32).

6-32. Overlay method.

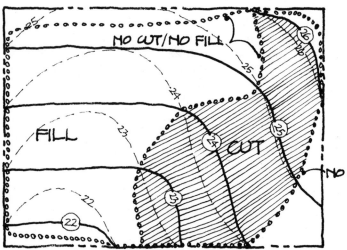

Average End Area

A third method of calculating cut and fill is the average end area method. The figures for this method are based on a series of vertical sections cut through the site, usually at a constant interval. The sum of the areas of two adjacent sections, multiplied by the linear distance between those areas and converted to cubic yards, gives an average cut or fill for the section, approximately the actual average. The closer together the sections are, the more accurate this method of calculation becomes. Figures derived from use of the average end area method tend to be high rather than low. As with the grid point method, the number of individual calculations required can be reduced by combining the figures that are common to two sections (figure 6-33).[4]

From the previous discussion of methods of calculating cut and fill, the importance of including both existing and proposed contours on the grading plan should be apparent. Unless both sets are shown, the contractor will have no way of determining how much earthmoving will be required and therefore will not be able to make accurate estimates of the cost, man-hours, or equipment needed to complete the job.

Grading is a compromise between desired appearance, economy, and function. Small sites must often sacrifice complex landforms to satisfy the primary function of getting the site to drain correctly. The logistics of moving large quantities of earth great distances may preclude major changes on large sites. Yet by paying careful attention to how the profile of the earth compares with the profile of the structures and enclosing elements, details, and plant materials, a sensitive, flowing, pleasing landscape can be created. The importance of aesthetically successful landscaping should not be underestimated. Awkward alignments of earth forms in relation to structural elements or plant masses can be just as disconcerting as a poor highway alignment, if not as dangerous.

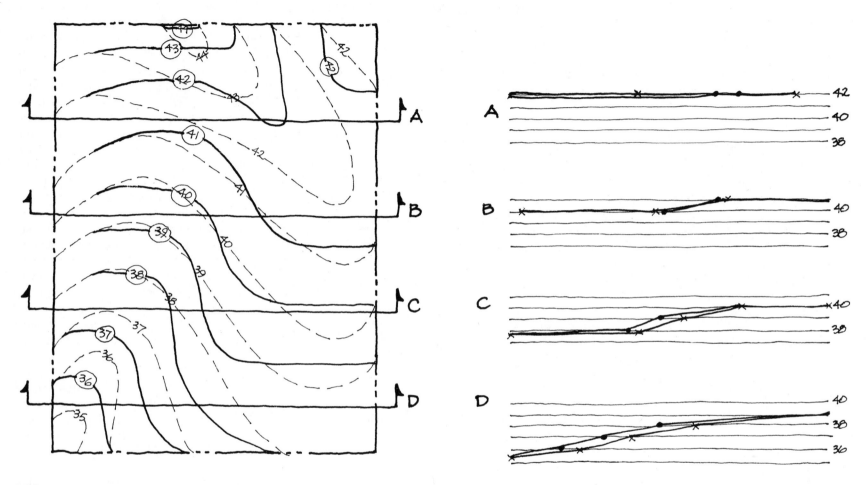

6-33. Average end area.

GRADING TO A SUBSURFACE DRAINAGE SYSTEM

Grading to provide positive surface drainage is the final step in the grading process on many sites, but it is only a start on those requiring subsurface drainage systems. All subsurface systems rely on a functioning surface system to receive the water; the location of drains and pipes is determined in relation to surface conditions.

Possible Legal Issues

Although the legal aspects of diverting or stopping the natural flow of water apply in the design of subsurface systems as they do in the design of surface drainage, the problems may not be as obvious—especially since many subsurface systems are installed primarily for the purpose of avoiding excessive drainage onto neighboring property. The sizing of a subsurface system, its manner of installation, and the designer's assessment of the data on which the system's layout is based can all become legal issues. An undersized system designed with minimal surface holding capacity can cause considerable damage in the areas it is intended to protect during infrequent, intense storms. Poorly installed pipes can go unnoticed for years, until the soil settles along its length and creates wet sinkholes.

Design Constraints

There are four main constraints on the installation of a subsurface storm drainage system. The first of these—the existence of a system in the near vicinity into which to tie the site system—is the most critical. This constraint is discussed in the section on grading (page 144).

A second constraint is that, in many cities, subsurface drainage is required in all commercial developments and new subdivisions; as a result, the developer may have to strike a compromise with the city's public works department to install adequate branch lines to the development. If building codes do require the use of a subsurface system, the designer has no choice but to comply and figure out how to arrange for most of the site drainage to take place underground—even though the natural pattern of the land might drain quite well across the surface.

A third constraint is that the existing topography of the site and surrounding area may either require or preclude the use of a subsurface system. If the area is extremely flat, the only way to achieve positive drainage around all site features may be to tip the grade artificially into inlets that carry the water off the site. A very steep site, on the other hand, may drain too well; a plan that surfaces areas that previously were permeable to runoff on such a site may result in water moving off the site with such speed that even small amounts cause damage to adjacent properties. In this situation construction becomes very expensive because the designer must avoid installing the system at so steep a pitch that excessive water speed builds up and causes turbulence and erosion at the discharge points. A subsurface system may be the only practical one under the circumstances.

Cost is the fourth constraint in the construction of subsurface systems. It relates directly to the difficulty of construction on the site, the length of run required, and the quantity and size of pipe and inlets. Current (1984) cost figures for a twelve-inch-diameter concrete pipe, for the material only, range from $3.50 to $5.65 per linear foot. Installed, that price rises to about $8.30 per linear foot. A forty-eight-inch-diameter reinforced concrete pipe costs $35 per linear foot, and $72 installed.[5] The design of the system should therefore reflect careful determinations of absolute minimum number of inlets needed to do the job of draining the site and of the most economical placement of pipe runs. On the other hand, whenever doubt exists about the ability of a subsurface system to handle the drainage adequately, the designer should plan to oversize rather than undersize the system.

SUBSURFACE DRAINAGE DEFINITIONS

The terminology used in subsurface draining should be understood before the designer attempts to work with storm drainage. Following are some basic terms and their definitions.

Runoff: The portion of water that reaches a drainage system after evaporation, plant intake, ponding, and absorption by the soil have occurred. Runoff is always less than the total amount of rainfall.

Watershed: A land area whose precipitation all drains into a single surface or subsurface system. A watershed's limits are defined by the land area's surrounding high points and ridge lines (figure 6-34).

Storm Sewer: The general name given to a man-made underground system of (surface) drain inlets and (subsurface) pipes, culverts, and head walls.

Drain Inlet: The structure through which surface water enters the subsurface system. Area drains (found in grass areas or parking lots), curb inlets (found at the intersections of streets), open culverts (found under the approach drives to farmsteads), and catch basins (found in the same locations as drains and curb inlets, but constructed differently) are all variations of the drain inlet (figure 6-35).

a.

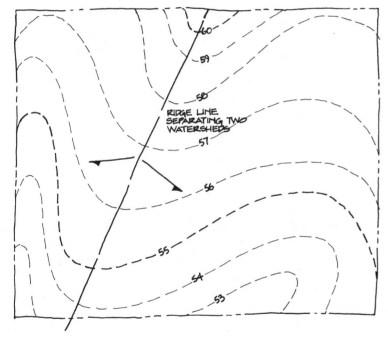

6-34. Watersheds.

b.

6-35. Drain inlets: a. one that is old-fashioned and dangerous; b. a new one at the bottom of an excessive slope.

Pipes: The subsurface conduits through which water is carried. Pipes may be composed of vitrified clay tile, concrete, reinforced concrete, polyvinyl chloride, cast iron, or corrugated metal.

Culvert: A pipe of large diameter designed to be laid in a surface drainage channel to allow water to pass under a necessary crossing. Usually, culverts are made of corrugated metal, but they can be made of concrete.

Head Walls: A construction providing an inlet point of a storm sewer system. Head walls are used as an alternative to a drain inlet where the surface swale feeds into the underground pipes. Usually concrete, with wing walls to channel the flow.

End Walls: A construction similar to head walls, with a concrete or surfaced splashway to help eliminate erosion (figure 6-36).

6-36. End wall.

Rim or *Top of Grate Elevation*: The elevation at an entrance point to a subsurface system, given in design plans for all drain and curb inlets, manholes, and other access structures.

Flow Line Elevation: The elevation of the inside bottom of the pipe through which the water will be running. Also called invert elevation.

Rational Formula: One of the formulas used to calculate the amount of water flowing in a drainage system for the purpose of sizing the pipes or channels. The formula is:

$$Q = Cia$$

where:

- Q = Quantity of water (in cubic feet per second)
- C = Coefficient of runoff dependent on the texture of the ground surface
- i = rainfall intensity in inches per hour
- a = area of the watershed in acres.

Design Storm: A quantity used to determine rainfall intensity (i) in the rational formula. Design storm is a subjective decision, based on judgment and past experience.

Time of Concentration: The time required (in minutes) for water to flow from the most remote point of the watershed to the outlet. Time of concentration combines overland flow with pipe flow in arriving at a final time-length.

THE SUBSURFACE DRAINAGE DESIGN PROCESS

The process of grading for a subsurface drainage system begins with the establishment of either the natural drainage patterns of the land or (in the case of a site that is to be extensively regraded) the proposed drainage patterns. This is done by dividing the site into watersheds along the ridge lines. Knowing the paths the water is likely to take as it flows perpendicularly down the slopes to the swales within each watershed allows the designer to assess alternative locations for proposed swale lines and low or gathering points. A trial surface grading plan identifies the ridge lines for the proposed changes.

Once the watershed areas have been identified, the surface area of each is calculated, using a planimeter for accuracy. The next step is to calculate the total quantity of runoff that must be handled through surface channels or a storm sewer system in each of the watersheds and in the site as a whole. The rational formula, described above, is relatively simple to apply and is accurate for watersheds of up to one hundred acres in area.

Values for the Rational Formula

Each material used in surfacing a site has a different rate of runoff and permeability, which substantially affects the amount of water that percolates into the soil and the amount that reaches the inlets and ultimately the pipes. The value of the variable C therefore changes from site to site. Since many watersheds are surfaced with more than one material, the coefficient of runoff for each material must be averaged in proportion to the amount of each material occurring on the site, in order to establish the variable C for the whole watershed. C depends on seasonal and daily fluctuations in climate and on variations in surfacing, so its assigned value is necessarily judgmental. Some values of C are:[6]

Rolling Woodland	0.30
Flat Pasture	0.30
Hilly Pasture	0.42
Flat Cultivated Land	0.50
Roofs	0.95
Concrete or Asphalt	0.95
Gravel	0.70
Parks and Lawn Areas	0.40

The calculation of rainfall intensity (i) is based on the maximum intensity of rainfall in inches per hour falling over the whole watershed with a selected frequency and duration (that is, on the rainfall for a chosen "design storm") and on the determined time of concentration of the runoff. Intensity can either be calculated (for very precise work) or found in tables used by many municipalities (figure 6-37). Since it is based on a subjectively chosen design storm, rainfall intensity, like the variable C, requires a judgment call by the designer. Storms of high intensity are usually of short duration and cover small land areas; however, these are potentially more damaging than longer, less intense storms because the speed at which the water falls does not allow it to be absorbed by the soil, causing more runoff.

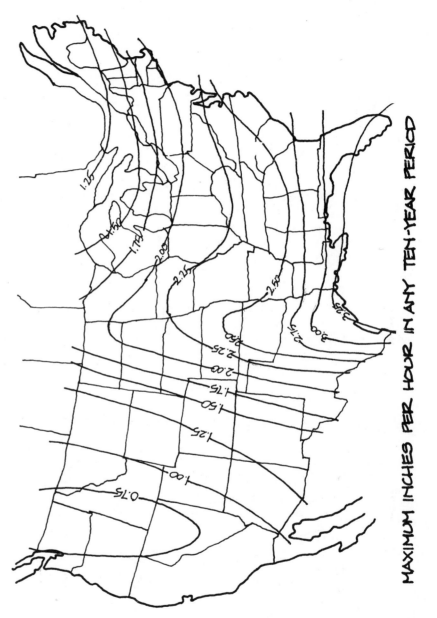

6-37. Rainfall intensities for the continental United States. The charts show predictable maximum one-hour rainfalls to be expected once in any given two-, five-, ten-, or twenty-five-year period. Adapted from Munson, *Construction Design for Landscape Architects.*

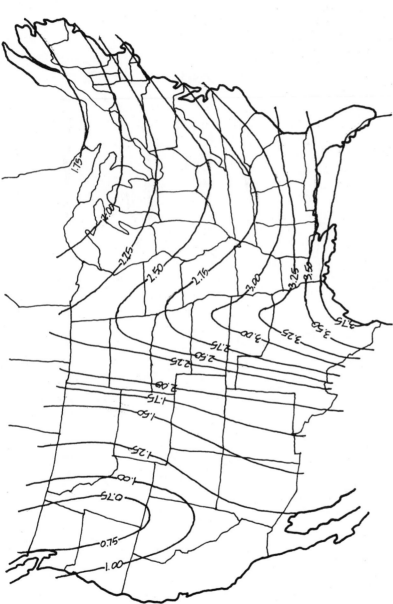

MAXIMUM INCHES PER HOUR IN ANY TWENTY-FIVE YEAR PERIOD

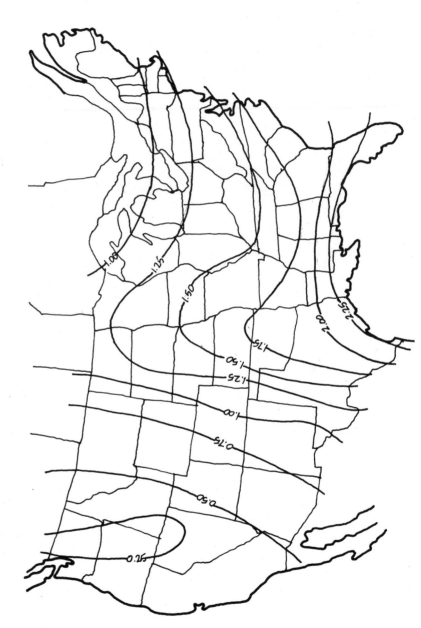

MAXIMUM INCHES PER HOUR IN ANY TWO-YEAR PERIOD

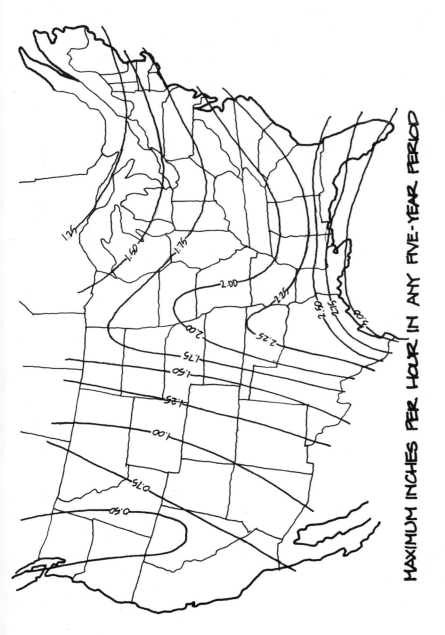

Time of concentration is the theoretical time runoff water takes to reach the outlet point of the system when released simultaneously from all points of the watershed. Generally, a fifteen- or twenty-minute overland flow time is an acceptable maximum for a workable design, although this can be calculated, too, using a chart similar to the one in figure 6-38. The pipe flow time can be obtained by dividing the length of the pipe by the average velocity of the water traveling through it.

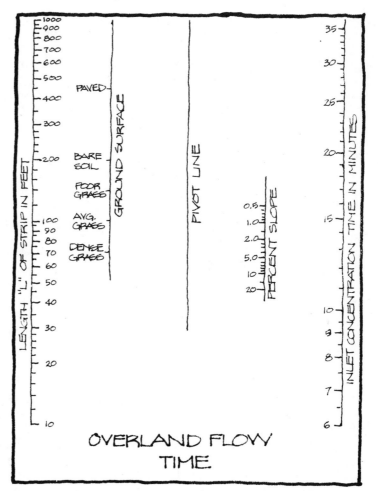

6-38. Overland flow time chart. Adapted from Seelye, *Data Book for Civil Engineers—Design*, Second Edition.

Sizing the Components

Once the quantity of runoff (Q) has been calculated for all watersheds on the site, the designer can determine more nearly exact locations for the storm sewer components and can determine the necessary pipe sizes. The flow of water across a site is cumulative; that is, more runoff will occur at the lowest point on the site than at the high end of the watershed. This means that progressively larger pipe is needed to handle the increasing volume of runoff and that the designer should begin calculating pipe sizes at the highest watershed and work downhill.

The following four factors are important in choosing pipe sizes: any pipe can empty into a pipe of the same size or larger, but not into a smaller pipe; maintaining a clean-out velocity of at least 2½ feet per second is desirable, but velocities of over 17 feet per second can cause destructive turbulence and erosion at the discharge point; building codes or freeze-thaw conditions often require a minimum cover over storm sewer pipes; and the flatter the slope of the pipe, and the larger the pipe, the greater the volume of water that can be carried.

The size of pipe required depends on the total volume of water that will be carried, its velocity, and the roughness or friction of the inside of the pipe. Since the velocity of the water in the pipe depends in part on the slope of the pipe, and since the slope is determined both by surface conditions that influence the elevations of the inlets and by flow lines of other pipes that must be met, velocity is a variable that can be changed to meet grade conditions, by raising or lowering the inlet or outlet end of the pipe.

If it is not possible to achieve a desired velocity and pipe size by changing surface elevations, the flow lines may be changed by either increasing or reducing the amount of cover over the pipe. On flat sites and in frost areas, however, a minimum depth must be maintained to avoid winter freeze-up, and changing the flow lines can become costly when pipes are deeply buried because deeper trenching and additional labor for installation and repair are then required.

By setting trial flow lines at both ends of the pipe, the designer sets a proposed slope. Pipe slopes are given in feet (of rise) per foot. Using the quantity Q and the trial slope, a pipe size can be chosen from a chart of different pipe sizes, such as the one shown in figure 6-39. If the intersection of quantity and slope falls between two pipe sizes, the larger pipe size should always be used.

After setting final rim and invert elevations and choosing a size for the first section of pipe, the designer works with the next section, computing the total volume of water that will be carried by that pipe by adding the volume of the pipe(s) that drain into it from higher watersheds, as well as the surface area of this particular watershed. Again the pipe-sizing chart is used, and the rim and flow lines of the pipe are adjusted to provide clean-out velocity without causing erosion or turbulence. Elevation adjustments and determinations of pipe size are made for the remainder of the system, and final elevations for rim and flow lines are assigned.

In summary, to design for either a surface or a subsurface drainage system, a trial grading plan is first developed, and high and low points are established. Existing grades that must be met and cannot be changed, proposed slopes for features that must meet certain slope criteria, and trial finish floor elevations determine the form this plan takes. Once positive drainage has been established around these features, the remainder of the contours are filled in, to drain either across the surface or into an underground system. The design of the underground system begins with calculation of the quantities of runoff for each of the watersheds outlined from the surface grading plan; it is designed from highest point to lowest, with each successive pipe carrying a larger amount of water than the one(s) above it.

SINGLE GRADING EXAMPLES

The development of a grading and drainage plan for most sites is composed of a number of small, rather simple components. Taken together, these can appear almost impossible to solve, but if the designer learns to handle each of these separately rather than all at once, the task of developing a workable grading plan will not seem nearly so overwhelming.

Plateau

A plateau grading scheme may be used on any site where additional emphasis on the height of the structure or where more drainage to a greater distance away from the building is desired.

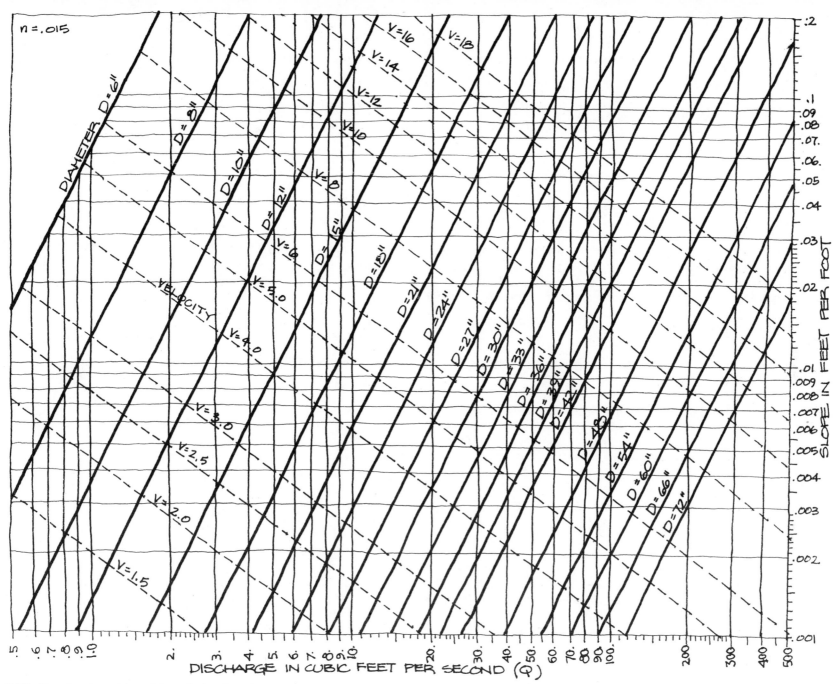

6-39. Pipe size chart. Adapted from Seelye, *Data Book for Civil Engineers—Design*, Second Edition.

The plateau may be so gradual as to be almost unnoticeable, or it may constitute a sharply defined architectural feature. Placing the structure on a plateau is one means of accomplishing surface drainage on a very flat site. By building a plateau, the designer is basically expanding the lower edges of the building outward to the toe of the grade feature, and the swale found in all positive surface drainage will be located beyond the toe (figure 6-40).

Swale

Swales created for drainage sometimes appear as distinct site features (figure 6-41). In such a case (and in the plateau example shown in figure 6-31), the finish floor elevation of the building must be set high enough to allow the swale contour to tie into the existing grade on the downhill side of the drainage, with enough room left uphill to match existing contours at the property line. Choosing a finish floor elevation that is too high or creating a plateau that extends too far out into the site will cause difficulties in matching the existing grades.

Pond

Where health regulations allow it and aesthetic considerations suggest its use, a pond might be created on the site as an alternative to taking the surface water off-site or underground to a storm sewer system. Grading a site to include a pond must provide a surface swale that will collect and channel as much of the surface water into the pond as possible, as well as a series of closed contours that represent the pond itself. An emergency spillway or outlet for excessive amounts of water is also desirable.

Since the water entering the pond will pass out through the other side unless it is contained, a dam of some sort must also be created. This can be a definite, engineered site feature (as it should be for large ponds) or only a slight rise in grade (for very small ponds). The soil characteristics on the site will help determine if and where a pond should be constructed (figure 6-42).

Subsurface System

The process to follow to design a storm sewer system was discussed on pages 168-72. The example shown in figure 6-43 assumes that the critical surface drainage is that across the paved surfaces. Many storm sewer systems are designed without providing much surface holding capacity at the drain inlets, meaning that during very heavy, fast rains much of the surface water will run by the inlet rather than into it. Holding capacity around a drain inlet is created in the same way that it is around a pond or low spot—by closing a contour or two around the inlet. However, in doing this, particularly in paved areas, the designer should make sure the slope remains constant, rather than becoming too steep at the inlet.

Terracing

Terracing is a way of making more of a steep site usable by flattening portions of it and separating these by even steeper slopes than were there before. A key to creating terraces that blend well with the surrounding grades is to make sure the toe of the terrace slope is smooth instead of abrupt. A terrace may extend several feet in the air or may involve a change of only a foot or two (figure 6-44).

Retaining Walls

In many cases, retaining walls may be aesthetically desirable as well as being desirable to compensate for grades that are so steep that even terracing cannot handle them. The structural strength of the retaining wall depends on the conditions of soil and surrounding slopes. A vast range of materials may be used for retaining wall construction, from dry-laid stone to railroad ties to concrete.

Retaining walls are often designed with a batter on the face; that is, instead of being exactly vertical, they are sloped back toward the earth they are retaining. This helps keep the wall from buckling from the weight of the earth.

When a contour line meets a retaining wall, it runs along the wall until the grade at the other end is reached. In plan, therefore, it may look as though contours are stacked on top of one another. The wall itself may slant gradually with the grade, or it may remain at a constant elevation. At the ends of the wall, the contours must return to natural grade. Too-steep slopes that become difficult to maintain often occur at these points, if the return to existing grade is overly abrupt (figure 6-45).

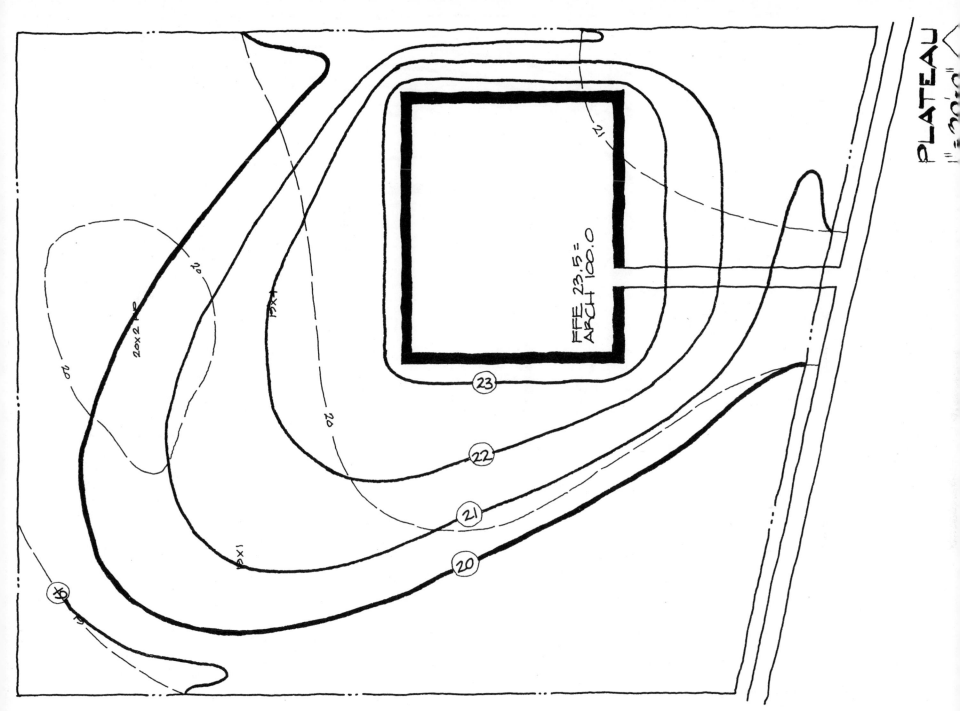

6-40. A plateau grading scheme.

Labels visible within the figure:

PLATEAU
1" = 20'-0"

FFE 23.5 =
ARCH 100.0

23

22

21

20

19

20

20

20

20×2 FP

12×1

14×1

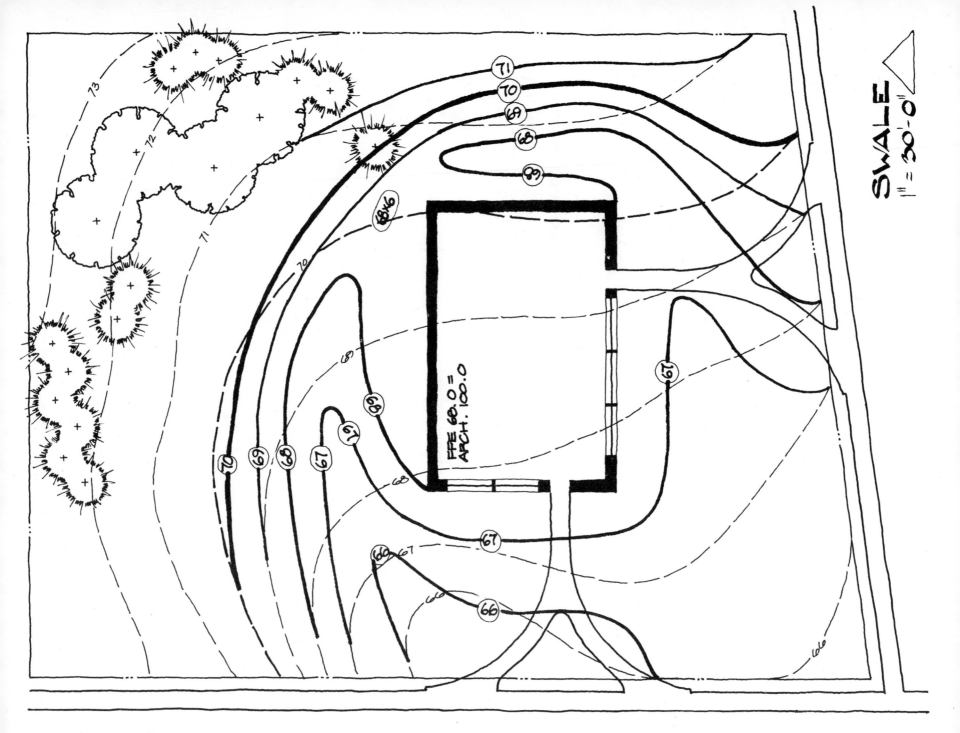

SWALE
1" = 30'-0"

FFE 60.0 = ARCH. 100.0

6-41. Grading to provide drainage swales around a building.

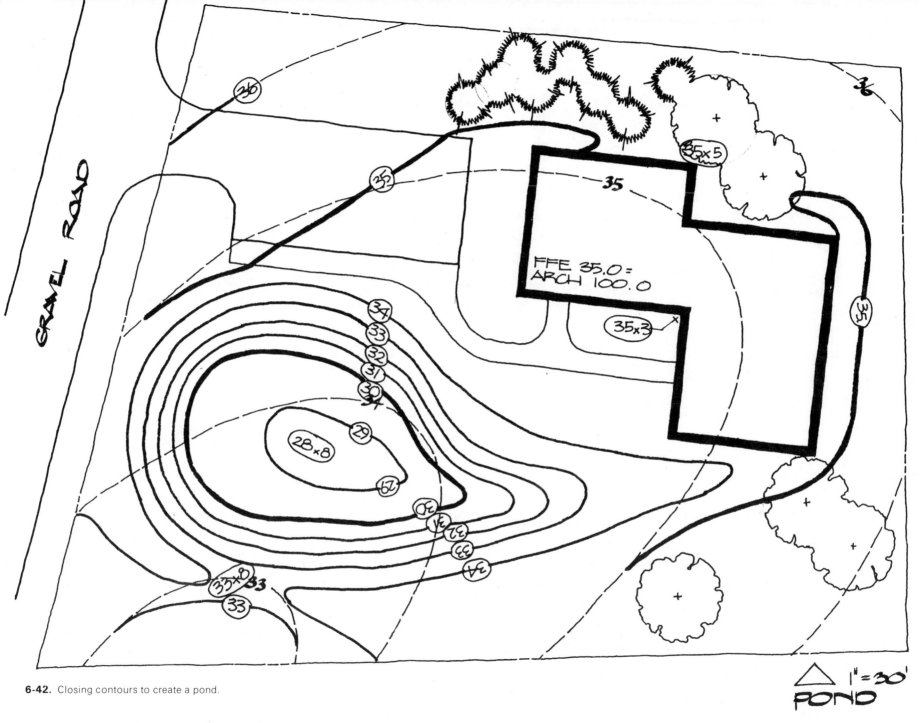

6-42. Closing contours to create a pond.

GRAVEL ROAD

FFE 35.0 = ARCH 100.0

POND

1" = 30'

Single Grading Examples **177**

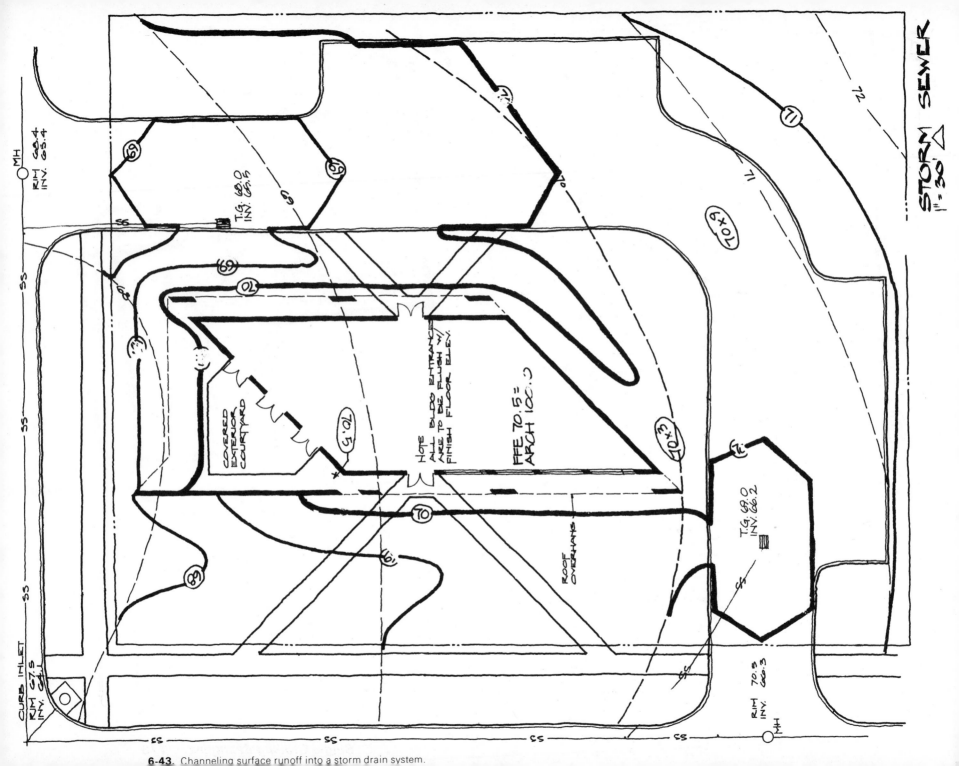

6-43. Channeling surface runoff into a storm drain system.

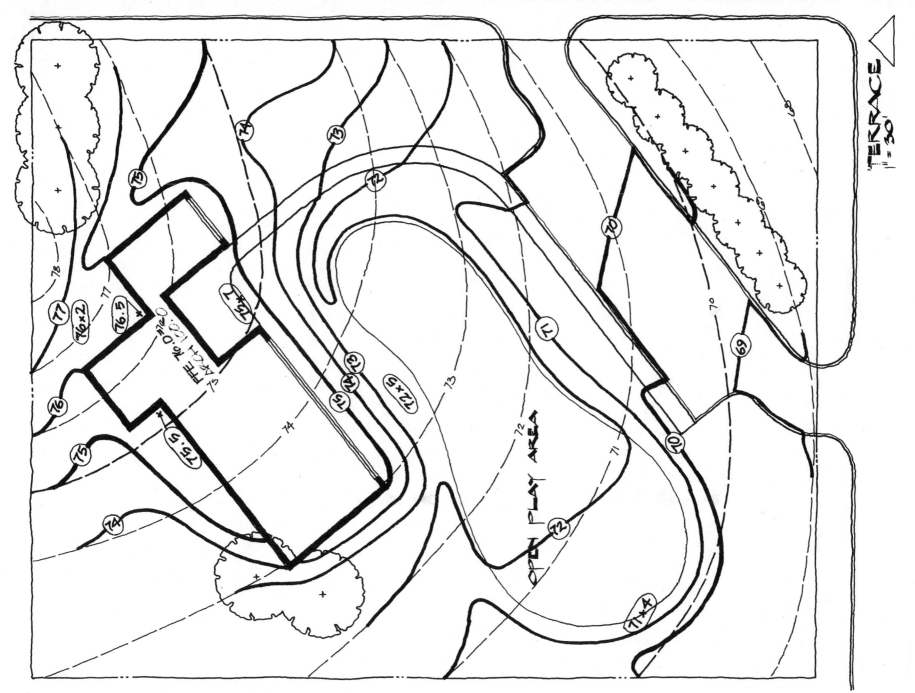

6-44. Terracing a steep slope to create level areas.

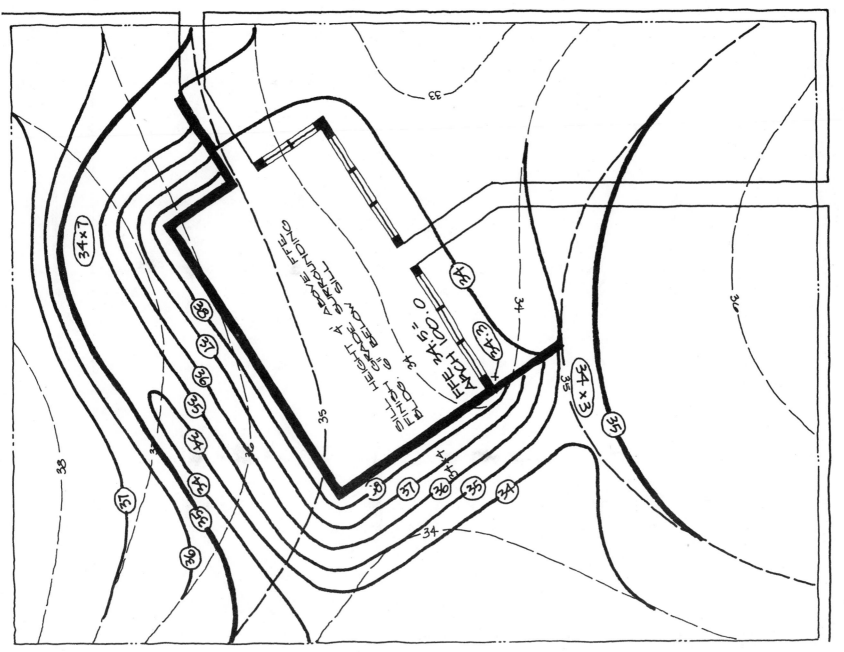

6-45. Using retaining walls to keep a berm away from the entrances.

Multilevel Structures

Grading for a multilevel structure that requires exterior access points on more than one level involves setting a finish floor elevation for each of those levels. Assuming that the change in elevation takes place inside, rather than by means of outside stairs, each level will have a different whole-number contour as its first contour away from the entrance. This means that the contour numbers occurring between one floor level and the next will lie between the two finish floor elevations, and the contour lines will die into the face of the building (figure 6-46).

Burying Structures

The use of berms to bury much of the foundation and walls of a structure is becoming increasingly popular as an energy conservation measure. Essentially, grading to bury a structure works the same way grading to place a structure on a plateau does: the berm extends the end of the building outward into the site, and the swale contour is located at that point. The height of the buried portion of the structure is a function of the amount of available space into which the landform is extended, the composition of the structure's materials, and the roof line (figure 6-47).

Freestanding Berms

Freestanding berms are used on many sites to screen undesirable views, to give a boost to small plant materials, and to provide climate control. Even though a berm is not composed of structural materials that can be damaged by water, positive drainage must still be provided around it to avoid erosion or wet spots. The designer therefore must establish a swale contour that, like the swale contour around a structure, matches existing drainage contours and patterns. The general form of the berm is decided by the designer, as is its desirable slope and top elevation. The slope and high point determine the breadth of the berm. A berm may be entirely freestanding, or it may be built against a wall (figure 6-48).

Parking Lots

There are many ways to grade a parking lot or other flat, paved surface. Common examples are sheet drainage, swale or crown drainage, and tipping the grade to one corner (from which water is allowed to run out of the lot and down a spillway). Multiple-bay parking lots may be graded to the centerline or gutters of each bay, forming a zigzag grading pattern that effectively drains each section without causing too much disturbance to any one area (figure 6-49).

Multiple-Structure Site

In grading a site that has more than one structure on it, the designer should follow the same basic procedure as for a single-structure site. Multiple-structure buildings can create multiple grading problems, however, if the site is either too flat or too steep or if there is insufficient room between structures to allow positive drainage around each. By carefully examining the existing conditions that cannot be changed and by establishing the finish floor elevations and swale contours for each of the structures, the designer can circumvent many problems (figure 6-50).

General Site

Once the designer understands the simpler components that may be encountered during grading and drainage and the variations that can occur in them as a result of different site conditions, the components can all be combined into a single project (figure 6-51). Following the analytical process described in this chapter and understanding where the site plan's flexibility lies and where the design is basically set are the essential parts of turning grading into a task that is readily accomplished.

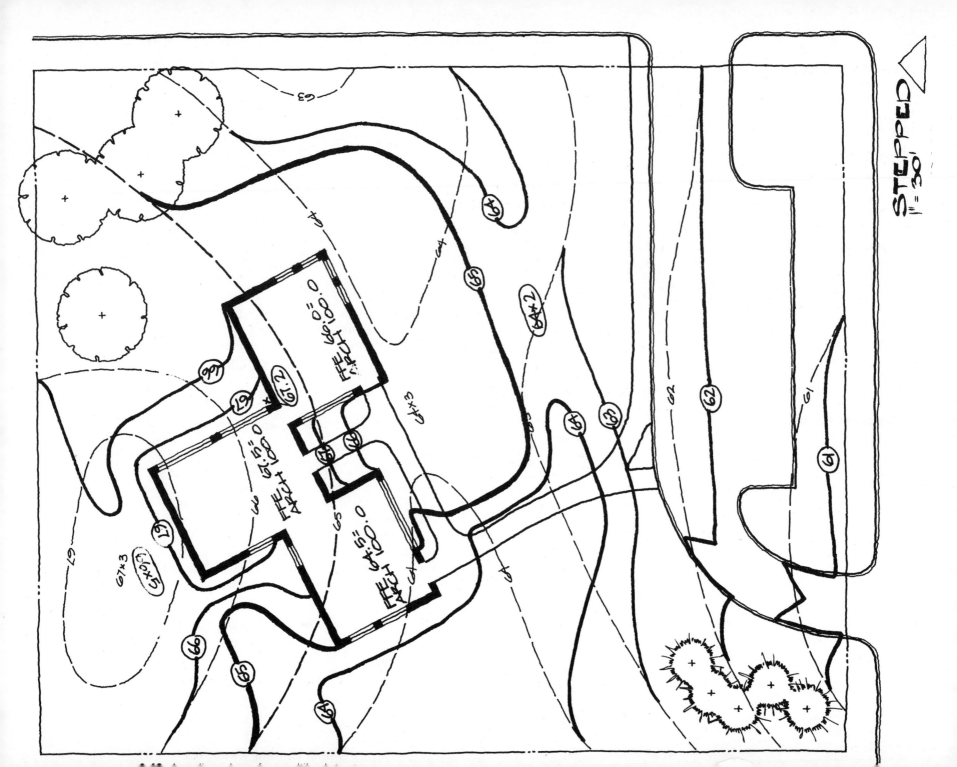

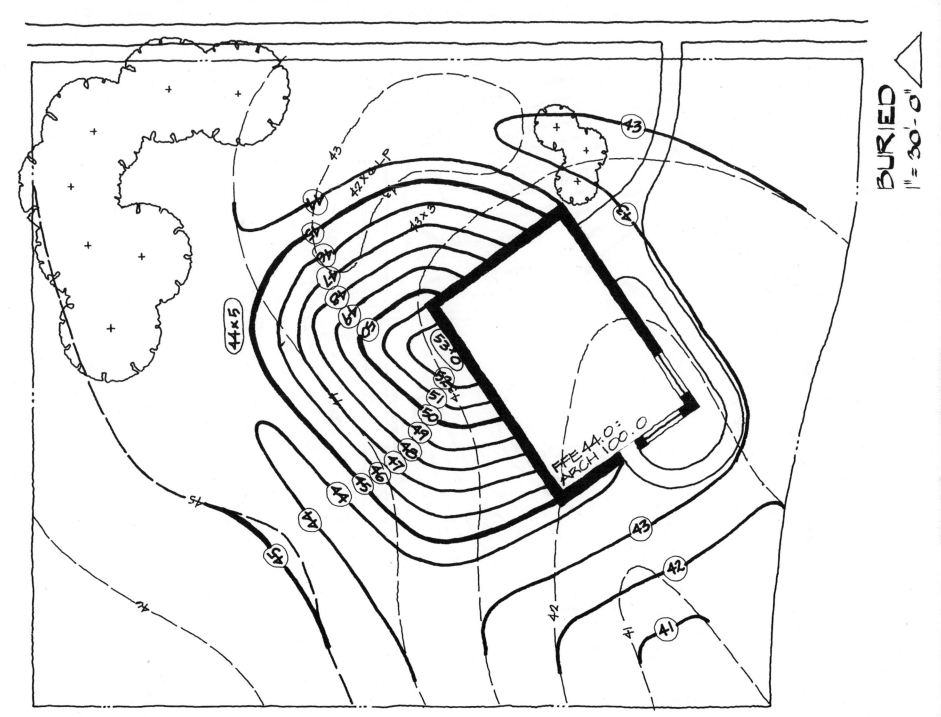

6-47. Using earth as insulation by piling it against the building.

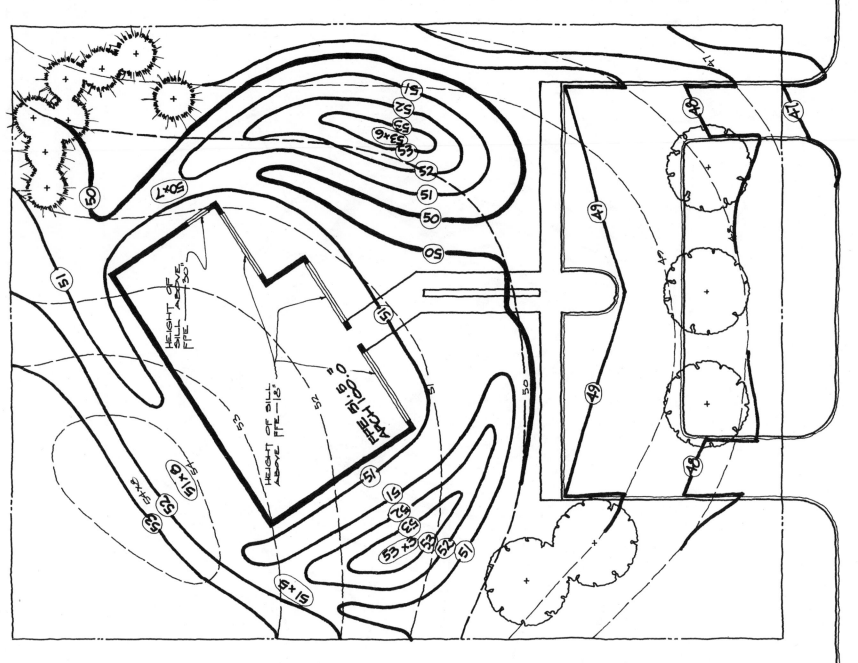

6-48. Earth berms provide a quick parking lot screen and help channel the wind.

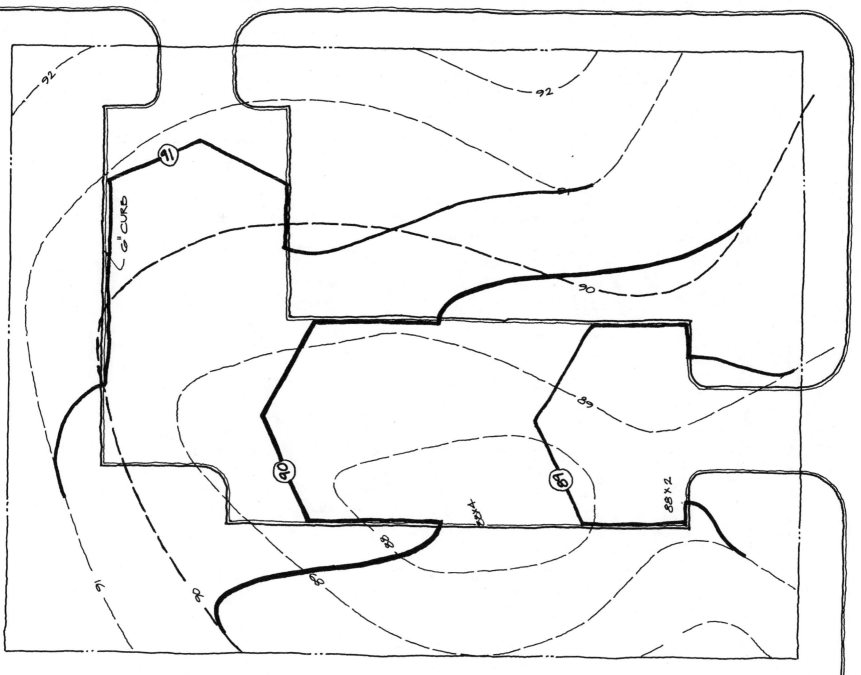

6-49. Grading a parking lot to a centerline swale.

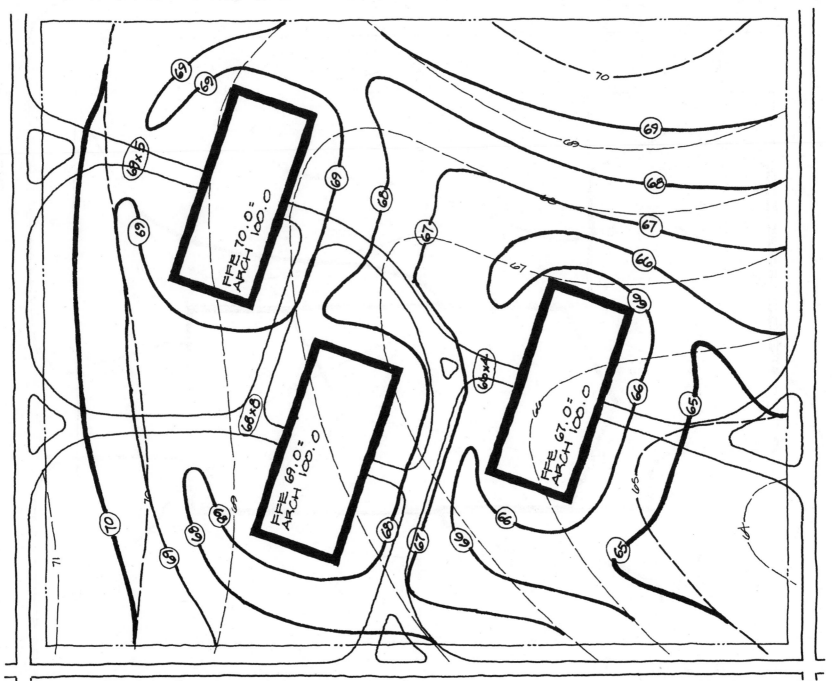

FFE 70.0 =
ARCH 100.0

FFE 69.0 =
ARCH 100.0

FFE 67.0 =
ARCH 100.0

6-50. Grading a site for multiple buildings.

MULTI BLDG.
1" = 40'

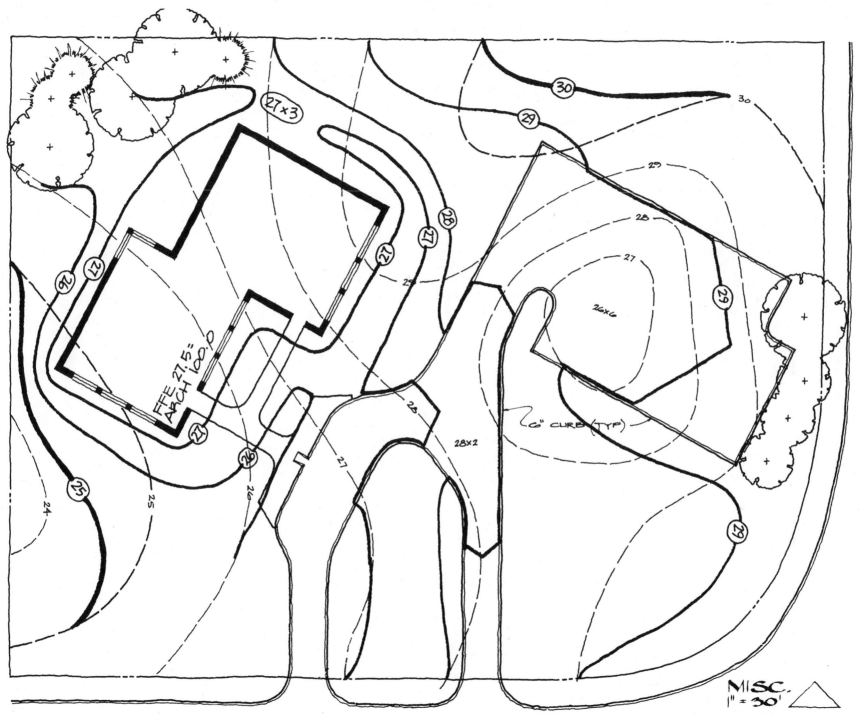

6-51. A typical site grading plan.

Within the figure:
- 27×3
- 30
- 29
- 30
- 29
- 28
- 27
- 26
- 27
- 28
- 29
- 27
- 26×6
- 26
- 25
- 25
- 24
- 28×2
- 26
- 27
- 28
- 29
- FFE 27.5 = ARCH 100.0
- 6" CURB (TYP)
- MISC.
- 1" = 30'

REFERENCES

1. Albe E. Munson, *Construction Design for Landscape Architects* (New York: McGraw-Hill, 1974).
2. Harlow C. Landphair, and Fred Klatt, Jr., *Landscape Architecture Construction* (New York: Elsevier North Holland, 1979).
3. American Society of Landscape Architects, *Landscape Architect's Handbook of Professional Practice* (McLean, Virginia: ASLA, 1972).
4. Kevin Lynch, *Site Planning*, 2nd ed. (Cambridge, Massachusetts: M.I.T. Press, 1971).
5. Robert Snow Means Co., *1983 Building Construction Cost Data* (Kingston, Massachusetts: Robert Snow Means Co., 1982), p. 43.
6. Iowa State University Department of Landscape Architecture, (Professor John M. Roberts), unpublished course notes, 1974.

Bibliography

American Society of Landscape Architects. *Landscape Architect's Handbook of Professional Practice*   McLean, Virginia: American Society of Landscape Architects, 1972.

Ashihara, Yoshinobu. *Exterior Design in Architecture*. Rev. ed. New York: Van Nostrand Reinhold, 1981.

Austin, Richard L. *Designing with Plants*. New York: Van Nostrand Reinhold, 1982.

Brinker, Russell C. *Elementary Surveying*. 5th ed. Scranton, Pennsylvania: International Textbook Co., 1969.

Eckbo, Garrett. *Urban Landscape Design*. New York: McGraw-Hill, 1964.

Hall, Edward T., *The Hidden Dimension*. New York: Doubleday & Co., 1966.

Halprin, Lawrence. *The RSVP Cycles: Creative Processes in the Human Environment*. New York: George Braziller, 1969.

Landphair, Harlow C., and Klatt, Fred, Jr. *Landscape Architecture Construction*. New York: Elsevier North Holland, Inc., 1979.

Lynch, Kevin. *Image of the City*. Cambridge, Massachusetts: Technology Press and Harvard University Press, 1960.

Lynch, Kevin. *Site Planning*. 2d ed. Cambridge, Massachusetts: M.I.T. Press, 1971.

McHarg, Ian L. *Design with Nature*. New York: Natural History Press, 1969.

Melaragno, Michele G. *Wind in Architecture and Environmental Design*. New York: Van Nostrand Reinhold, 1982.

Munson, Albe, E. *Construction Design for Landscape Architects*. New York: McGraw-Hill, 1974.

Olgyay, Victor. *Design with Climate*. Princeton, New Jersey: Princeton University Press, 1973.

Relph, E. *Place and Placelessness*. London: Pion Limited, 1976.

Robinette, Gary O., ed. *Landscape Planning for Energy Conservation*. American Society of Landscape Architects Foundation. Reston, Virginia: Environmental Design Press, 1977.

Rubenstein, Harvey M. *A Guide to Site and Environmental Planning*. New York: John Wiley & Sons, 1969.

Seelye, Elwin E. *Data Book for Civil Engineers: Volume I, Design*, 3d ed. rev. New York: John Wiley & Sons, 1960.

Simonds, John Ormsbee. *Landscape Architecture*. New York: McGraw-Hill, 1961.

Wyman, Donald, and Prendergast, Curtis. *Easy Gardens*. Alexandria, Virginia: Time-Life Books, 1978.

Wyman, Donald. *Shrubs and Vines for American Gardens*. New York: Macmillan, 1958.

Wyman, Donald. *Trees for American Gardens*. New York: Macmillan, 1959.

Zion, Robert L. *Trees for Architecture and the Landscape*. New York: Reinhold Book Corp., 1968.

Index